Wireless Networks

Series Editor

Xuemin Sherman Shen, University of Waterloo, Waterloo, Canada

The purpose of Springer's Wireless Networks book series is to establish the state of the art and set the course for future research and development in wireless communication networks. The scope of this series includes not only all aspects of wireless networks (including cellular networks, WiFi, sensor networks, and vehicular networks), but related areas such as cloud computing and big data. The series serves as a central source of references for wireless networks research and development. It aims to publish thorough and cohesive overviews on specific topics in wireless networks, as well as works that are larger in scope than survey articles and that contain more detailed background information. The series also provides coverage of advanced and timely topics worthy of monographs, contributed volumes, textbooks and handbooks.

Yuhan Kang • Hao Gao • Zhu Han

Mean Field Guided Machine Learning

Springer

Yuhan Kang
Weichai America Corporation
Houston, TX, USA

Hao Gao
Samsung
Houston, TX, USA

Zhu Han
The University of Houston
Houston, TX, USA

ISSN 2366-1186 ISSN 2366-1445 (electronic)
Wireless Networks
ISBN 978-3-031-91858-2 ISBN 978-3-031-91859-9 (eBook)
https://doi.org/10.1007/978-3-031-91859-9

This Springer imprint is published by the registered company Springer Nature Switzerland AG
The registered company address is: Gewerbestrasse 11, 6330 Cham, Switzerland

Preface

Machine learning (ML) has revolutionized numerous industries, transforming how we approach complex problems by leveraging vast amounts of data and advanced computational techniques. However, as powerful as these technologies have become, they still face significant challenges, particularly in high-dimensional spaces and multi-agent systems. The need to manage complexity, enhance scalability, and maintain robustness in increasingly sophisticated applications has never been greater.

In this book, we explore the intersection of Mean Field Game (MFG) theory and machine learning, an innovative approach that holds promise for addressing these pressing challenges. MFG theory, which models the collective behavior of large populations of interacting agents, provides powerful mathematical tools for simplifying and optimizing complex systems. By integrating MFG theory with ML, we can enhance the efficiency, scalability, and robustness of machine learning models, opening new avenues for research and application development.

This book is the culmination of extensive research and collaboration among the authors, who bring diverse expertise in machine learning, game theory, and computational mathematics. We have endeavored to present a comprehensive guide that bridges the gap between these two fields, offering both theoretical insights and practical applications. Our aim is to equip researchers, practitioners, and students with the knowledge and tools needed to harness the full potential of MFG-enhanced machine learning.

Throughout the book, we delve into the foundational concepts of MFG theory and ML, survey recent advances, and discuss the challenges and motivations driving this integration. We provide detailed examples and case studies to illustrate how MFG can be applied to enhance various machine learning techniques, from supervised and unsupervised learning to reinforcement learning and federated learning.

We hope this book will serve as a valuable resource for those seeking to understand and leverage the synergy between MFG theory and machine learning. The journey of exploring this intersection is both challenging and rewarding, and we are excited to share our insights and findings with the broader scientific community.

We would like to express our gratitude to our colleagues, mentors, and the broader research community for their support and contributions. Special thanks go to our families for their unwavering encouragement and understanding throughout this endeavor.

We would like to express our deepest gratitude to everyone who has contributed to the completion of this book. First and foremost, we extend our heartfelt thanks to our co-authors, whose dedication, expertise, and collaboration have been instrumental in bringing this work to fruition.

We also extend our gratitude to our colleagues and mentors who have provided invaluable feedback and guidance throughout the development of this book. Your insights and encouragement have been vital in shaping the direction and quality of our work.

We are immensely grateful for the financial support provided by various institutions. Special thanks to the National Science Foundation (NSF), Amazon, Toyota, the US Department of Transportation, and the Japan Science and Technology Agency. Your generous funding and resources have been crucial in advancing our research and making this book possible.

Lastly, we want to express our deepest appreciation to our families and friends for their unwavering support, patience, and understanding. Your encouragement has been a source of strength and motivation throughout this journey.

Thank you all for your contributions and support. We are excited to share the results of our efforts with the broader scientific community and look forward to the advancements and innovations that will emerge from this work.

Houston, TX, USA
May 28, 2024

Yuhan Kang
Hao Gao
Zhu Han

Contents

Acronyms

ML	Machine Learning
MFG	Mean Field Game
NSF	National Science Foundation
PDEs	Partial Differential Equations
HJB	Hamilton-Jacobi-Bellman
FPK	Fokker-Planck-Kolmogorov
AI	Artificial Intelligence
PCA	Principal Component Analysis
RL	Reinforcement Learning
FL	Federated Learning
DQN	Deep Q-Networks
FA	Federated Averaging
ODEs	Ordinary Differential Equations
LLMs	Large Language Models
NLP	Natural Language Processing
ViTs	Vision Transformers
CNNs	Convolutional Neural Networks
BERT	Bidirectional Encoder Representations from Transformers
GANs	Generative Adversarial Networks
SSL	Self-Supervised Learning
AutoML	Automated Machine Learning
NE	Nash Equilibrium
SVM	Support Vector Machines
RBF	Radial Basis Function
ResNets	Residual Networks
MDP	Markov Decision Process
NNs	Neural Networks
SNs	Social Networks
HK Model	Hegselmann-Krause Model
W-GAN	Wasserstein Generative Adversarial Neural Network
W-2	Wasserstein-2

KL Divergence	Kullback–Leibler Divergence
DDPG	Deep Deterministic Policy Gradient
DRL	Deep Reinforcement Learning
MFG-Q	Mean Field Guided Q-Value
MEC	Multi-access Edge Computing
TD	Temporal Difference
MFED	Mean Field Evolutionary Dynamics
MFRL	Mean Field Reinforcement Learning
LEO	Low-Earth-Orbit
SatComs	Satellite Communications
MFEv	Mean-Field-Evolutionary
IoT	Internet-of-Thing
RIS	Reconfigurable Intelligent Surface
IP	Integer Programming
RD	Replicator Dynamics
DNNs	Deep Neural Networks
NAS	Neural Architecture Search
EC	Evolutionary Computation
ENAS	Evolutionary Computation-Based Neural Architecture Search Algorithms
DRE	Driving Range Estimation
BEV	Battery Electric Vehicle
SoC	State of Charge
V2V	Vehicle to Vehicle
SGD	Stochastic Gradient Descent
LSTM Network	Long Short-Term Memory Network

Chapter 1
Overview of Mean Field Theory and Machine Learning

In recent years, the fields of mean field game (MFG) theory and machine learning (ML) have seen remarkable advancements, each transforming the landscape of their respective domains. MFG theory, initially developed to tackle optimal control problems involving large populations of interacting agents, provides a powerful framework for modeling complex systems where individual behaviors aggregate to influence the collective dynamics. On the other hand, ML, a subset of artificial intelligence, focuses on developing algorithms that enable systems to learn from data, identify patterns, and make informed decisions. The convergence of these two fields presents an exciting frontier in computational science, offering innovative solutions to challenges inherent in large-scale, multi-agent environments. This book aims to bridge the gap between MFG theory and ML, exploring how the principles of MFG can enhance ML techniques and vice versa. Through a comprehensive examination of theoretical foundations, recent advancements, and practical applications, this book provides insights into the synergistic potential of integrating MFG and ML to address complex real-world problems efficiently and effectively.

1.1 Background and Prerequisites

1.1.1 Background of Mean Field Game Theory

MFG theory was first introduced by Jean-Michel Lasry and Pierre-Louis Lions in the mid-2000s [117]. The theory was developed to address optimal control problems involving a large number of indistinguishable and rational players. These players interact with each other, and the collective behavior of the population can be described using the concept of a "mean field." The mean field approximates the aggregate effect of all individuals, significantly simplifying the analysis and computation of large-scale dynamic systems.

Y. Kang et al., *Mean Field Guided Machine Learning*, Wireless Networks,
https://doi.org/10.1007/978-3-031-91859-9_1

MFG theory emerged as a response to the increasing complexity of modeling systems with numerous interacting agents. Traditional game theory struggles with the "curse of dimensionality" when applied to such systems, making it infeasible to analyze and solve problems with a vast number of players. MFG theory overcomes this by considering the limit as the number of players tends to infinity, allowing for a tractable and insightful analysis. On the other hand, MFG theory is grounded in differential equations, particularly partial differential equations (PDEs), which describe the evolution of the mean field over time. The key components of MFG include the Hamilton-Jacobi-Bellman (HJB) equation, which determines the optimal strategy for individual players, and the Fokker-Planck-Kolmogorov (FPK) equation, which describes the distribution of the states of the players over time. By iteratively solving these equations, one can find the Nash Equilibrium for the system, where no player has an incentive to deviate from their optimal strategy given the mean field.

1.1.1.1 Key Concepts in Mean Field Game Theory

To provide a clearer understanding of Mean Field Games (MFGs), some essential concepts are introduced below:

- *Mean Field Interaction:* The core idea of MFG theory is that each player is influenced by the average effect of the other players' states and actions rather than by individual interactions. This mean field represents the collective influence of the entire population, making it easier to analyze and predict the behavior of the system.
- *Continuum of Players:* Instead of dealing with a finite number of players, MFG theory models a continuum of players. This abstraction allows for a more straightforward mathematical treatment and leads to results that approximate the behavior of large but finite systems.
- *Optimal Control and Equilibrium:* In MFG theory, players aim to optimize their individual strategies based on the mean field. The equilibrium state is reached when no player has an incentive to deviate from their optimal strategy, given the mean field. This equilibrium is characterized by a stable distribution of players' states and actions.

1.1.1.2 Applications of Mean Field Games

MFG theory has found applications in various domains such as economics, finance, network theory, and engineering. Some application examples are given below.

- *Economics:* In economics, MFG helps model market dynamics where individual agents' actions collectively influence market prices. For instance, in a financial market, traders' buying and selling decisions impact the overall market behavior.

MFG can model this by treating each trader as a player whose strategy affects and is affected by the mean field representing market conditions.

- *Finance:* In finance, MFG is used for portfolio optimization and risk management. Investors' decisions on asset allocation influence the market, and MFG provides a framework to determine optimal investment strategies by considering the collective behavior of all investors.
- *Network Theory:* In network theory, MFG can be used to optimize the performance of large communication networks by modeling the interactions among numerous nodes. For example, in a wireless communication network, each device's transmission power affects the overall network performance. MFG can help design protocols that optimize network efficiency and resource allocation.
- *Engineering:* In engineering, MFG is applied to the control of large-scale systems such as power grids and transportation networks. By modeling the interactions of numerous components, MFG provides insights into optimal control strategies that enhance system performance and reliability.
- *Social Sciences:* In social sciences, MFG models crowd behavior and social dynamics. For example, in modeling evacuation scenarios, MFG can predict how individuals' decisions to move towards exits impact the overall evacuation process, helping in designing better emergency response plans.

For more details of MFGs, please find in Chap. 2.

1.1.2 Background of Machine Learning

ML is a subset of artificial intelligence (AI) that focuses on the development of algorithms that enable computers to learn from and make predictions or decisions based on data. The roots of ML can be traced back to the mid-twentieth century with the advent of statistical modeling and neural networks. However, significant advancements in computational power, availability of large datasets, and development of sophisticated algorithms have propelled ML into the forefront of technological innovation in recent years.

ML techniques can be broadly categorized into supervised learning, unsupervised learning, reinforcement learning, and federated learning:

- *Supervised Learning*: In supervised learning, the algorithm learns from labeled training data, making predictions or classifications based on input-output pairs. Techniques such as linear regression, support vector machines, and neural networks are widely used in this domain.
- Unsupervised Learning: Unsupervised learning involves finding hidden patterns or intrinsic structures in input data without labeled responses. Clustering and dimensionality reduction techniques, such as k-means clustering and principal component analysis (PCA), are common unsupervised learning methods.
- Reinforcement Learning (RL): RL focuses on training agents to make sequences of decisions by maximizing cumulative rewards received from interacting with

an environment. Notable advancements include Deep Q-Networks (DQN) and Policy Gradient methods, which have achieved remarkable success in fields like game playing and robotics.

- Federated Learning (FL): FL is a decentralized approach to ML that enables model training across multiple devices without sharing raw data. This method preserves privacy and security, making it ideal for applications in healthcare, finance, and IoT. Techniques like Federated Averaging (FedAvg) exemplify this approach.

The development of machine learning (ML) has surged in recent years. Meta-learning, large language models, generative AI, and transformers are just a few examples of the advancements in the current ML landscape. For more details on recent advances in ML, please refer to Chap. 1.2.2.

The connection between MFG theory and ML is particularly tight and synergistic. The agents in MFG can be directly mapped to various elements within ML frameworks: they can represent data samples in a dataset, neurons in a neural network, agents in reinforcement learning, or local clients in federated learning. This versatile representation allows MFG to be applied across different ML paradigms, leveraging its capability to model the collective behavior of large populations. In supervised learning, for instance, data samples can be treated as agents whose interactions affect the overall learning process. In neural networks, neurons can be seen as agents that collectively learn to optimize a loss function. In reinforcement learning, MFG provides a framework to model and optimize the interactions among multiple learning agents. Similarly, in federated learning, local clients can be considered as agents that independently learn and contribute to a global model. By integrating MFG with ML, we can enhance the scalability, efficiency, and robustness of ML models, making them more adept at handling large-scale, distributed, and dynamic environments.

1.1.3 Prerequisites for This Book

To fully appreciate the integration of MFG theory and ML, readers should have a solid understanding of several foundational concepts. Mastery of these requirements is crucial not only for comprehending the content of this book but also for grasping future advancements in these techniques.

- Differential Equations: Knowledge of ordinary differential equations (ODEs) and partial differential equations (PDEs) is essential, as MFG theory relies heavily on these mathematical tools to model the dynamics of large populations and their interactions.
- Probability and Statistics: A strong grasp of probability theory and statistical methods is crucial for understanding the stochastic elements of MFG and the statistical learning techniques used in ML.

- Game Theory: Familiarity with the basic principles of game theory, including concepts like Nash Equilibrium and optimal strategies, will provide a deeper understanding of how MFG models the interactions among rational agents.
- Machine Learning Algorithms: Proficiency in key ML algorithms, including supervised, unsupervised, reinforcement, and federated learning techniques, will enable readers to appreciate how MFG can enhance and optimize these methods.
- Computational Skills: Practical experience with computational tools and programming languages, such as Python and MATLAB, will be beneficial for implementing and experimenting with MFG and ML models.

Overall, this book is designed for advanced undergraduate students, graduate students, and researchers in the fields of computer science, applied mathematics, or related disciplines, who are interested in the fusion of game theory and machine learning. Practitioners in the machine learning field who are keen on exploring new horizons and incorporating MFG into their toolkits may also find this book invaluable.

Despite these prerequisites, we've aimed to make the book as comprehensive and self-contained as possible. Key concepts are explained with a detailed yet concise approach, and each chapter builds upon the knowledge imparted in the previous ones, ensuring a smooth and enlightening learning journey for the readers.

1.2 Survey of Recent Advance

We provide a comprehensive overview of the latest advancements in Mean Field Games (MFG) and Machine Learning (ML) in this section. We aim to highlight the significant developments that have shaped these fields in recent years and explore how their integration can address current challenges and enhance the capabilities of ML systems. By examining both the theoretical innovations and practical applications, this chapter sets the stage for understanding the synergistic potential of combining MFG theory with advanced ML techniques.

1.2.1 Recent Advances in Mean Field Games

Since its inception by Jean-Michel Lasry and Pierre-Louis Lions in the mid-2000s, Mean Field Game (MFG) theory has rapidly evolved and expanded, finding applications in various fields such as economics, finance, network theory, and engineering. The foundational concept of MFG is to approximate the collective behavior of a large population of interacting agents through a "mean field," significantly simplifying the analysis and computation of complex systems.

1.2.1.1 Advancements in Theoretical Frameworks of MFGs

Recent years have seen significant advancements in the theoretical frameworks of MFG. Researchers have extended the classical MFG models to include various complexities such as:

- *Common Noise and Stochastic Environments:* Incorporating common noise into MFG models allows for the consideration of randomness that affects all agents simultaneously, enhancing the realism of the models in economic and financial contexts [27].
- *Heterogeneous Agents:* Developing MFG models that account for heterogeneity among agents, such as differences in preferences, risk tolerance, or strategies, has been a major focus. This extension allows for more accurate modeling of real-world systems where agents are rarely identical [1].
- *Mean Field Control:* Advances in mean field control have provided new insights into optimal control strategies in large-scale systems. This includes the development of mean field-type control frameworks where the control actions of agents are influenced by the mean field [15].

MFG theory has been applied to numerous domains, leading to innovative solutions and new research directions:

- *Economics and Finance:* In economics, MFG models are used to analyze market dynamics, including price formation and the impact of individual trading strategies on the market. In finance, MFG helps in understanding the behavior of a large number of investors and their influence on asset prices and market volatility [26].
- *Network Theory:* MFG has been utilized to optimize the performance of communication networks by modeling the interactions among nodes. This includes applications in wireless communication, where the goal is to efficiently allocate resources and manage interference among a large number of devices [93].
- *Engineering and Control Systems:* In engineering, MFG theory aids in controlling large-scale systems such as power grids, edge computing, mobile robots, and other transportation networks. By understanding the aggregate behavior of numerous interacting components, MFG provides strategies for efficient resource allocation and system stability [22, 103].

1.2.2 Recent Advances in Machine Learning

In the vast tapestry of modern computational technologies, ML is emerging as a decisive player, spearheading the fourth industrial revolution. The development of machine learning over the last few decades has been meteoric, penetrating and transforming myriad sectors, spanning from healthcare [63, 76] to finance [185], transportation [17], marketing [115], and beyond. It's a revolutionary technological wave characterized by the fine craft of instructing machines to learn from data, draw inferences, and make decisions or predictions, thereby simulating human intelligence, but at a scale and speed far beyond human capabilities.

ML operates by identifying patterns in data, developing models that explain these patterns, and using these models to predict future data or other outcomes of interest. Successful ML tasks have already been achieved in the industry field including predictive maintenance, quality control and anomaly detection, supply chain optimization, energy load forecasting, autonomous vehicles [168], fraud detection [157], etc. The ability to perform these tasks effectively often depends on large volumes of high-quality data and substantial computational power. Machine Learning methods can be broadly categorized into reinforcement learning, supervised learning, and unsupervised learning, each with its unique set of strengths and applications.

- *Reinforcement Learning:* A significant development in ML is Reinforcement Learning (RL). RL focuses on teaching agents to make sequential decisions in dynamic environments. Inspired by how humans and animals learn through trial and error, RL algorithms enable an agent to learn optimal actions based on feedback received from the environment. Unlike traditional ML, where models are trained on labeled or unlabeled data, RL operates in a different paradigm. In RL, an agent interacts with an environment, observes its state, takes actions, and receives feedback in the form of rewards or penalties. The goal of the agent is to learn a policy (i.e., a mapping of states to actions) that maximizes the cumulative rewards over time. Recent advancements in RL include Deep Q-Networks (DQN), Policy Gradient methods, and the AlphaGo series, which have demonstrated the potential of RL in complex, high-dimensional environments.
- *Transformers:* Recent years have also seen remarkable progress in neural network architectures, particularly with the advent of Transformers and Large Language Models (LLMs). Transformers, introduced in [187] by Vaswani et al., have revolutionized natural language processing (NLP). These models leverage self-attention mechanisms to capture contextual relationships in text data, leading to state-of-the-art performance in various NLP tasks such as translation, summarization, and sentiment analysis. The Transformer architecture has also been extended to other domains, such as image processing with Vision Transformers (ViTs), which have shown competitive performance with traditional Convolutional Neural Networks (CNNs).
- *Large Language Models:* LLMs, like OpenAI's GPT-3 and GPT-4, have pushed the boundaries even further, enabling sophisticated text generation, question-answering, and conversational agents. These models are capable of understanding and generating human-like text, opening up new possibilities for applications in customer service, content creation, and more. Recent advances also include BERT (Bidirectional Encoder Representations from Transformers) and its successors, which have set new benchmarks in various NLP tasks by pre-training on large text corpora and fine-tuning on specific tasks.

- *Generative Adversarial Networks:* GANs have also emerged as a powerful tool for generative modeling, capable of creating highly realistic synthetic data [43]. By pitting two neural networks–the generator and the discriminator–against each other in a competitive framework, GANs have achieved impressive results in image synthesis, data augmentation, and generating high-resolution images from low-resolution inputs. Recent advancements in GANs include StyleGAN, which allows for fine-grained control over the generated images, and BigGAN, which scales up the GAN architecture to produce higher-quality images.
- *Federated Learning:* With the ever-increasing development of ML, privacy breaches have raised significant concerns. In traditional ML, data is collected from various sources and aggregated in a central server for model training. However, this approach may not be feasible or desirable in scenarios involving mobile devices, Internet of Things (IoT) devices, or sensitive personal data. This motivated the development of Federated Learning (FL), a collaborative form of ML where models are trained across multiple decentralized edge devices or servers holding local data samples, without exchanging them. FL is beneficial in scenarios where privacy, data security, and data locality are paramount. It brings the algorithms to the data rather than the traditional approach of bringing data to the algorithms. FL has significant applications in healthcare, finance, and IoT, where data privacy and security are critical. Notable advancements in FL include Federated Averaging (FedAvg) and privacy-preserving techniques like Differential Privacy and Secure Multiparty Computation.
- *Self-Supervised Learning (SSL):* Another recent technique gaining traction is Self-Supervised Learning (SSL), where models learn useful representations from unlabeled data by solving pretext tasks [99, 137]. This approach reduces the reliance on large labeled datasets, which are often expensive and time-consuming to obtain. SSL has been successfully applied in NLP with models like BERT and GPT, and in computer vision with models like SimCLR and MoCo.
- *Auto Machine Learning (AutoML):* The rise of AutoML has also simplified the process of model selection and hyperparameter tuning. AutoML frameworks, such as Google's AutoML, Auto-Keras, and H2O.ai [86, 100, 118], automate the end-to-end process of applying ML to real-world problems, making it accessible to non-experts and enhancing productivity for experienced practitioners.

MFG theory, a relatively new branch of game theory, has gained considerable attention in recent years due to its potential applications in various fields, including economics, wireless communications, transportation, and social sciences [42, 93, 117]. Mean field games provide a framework to model large populations of interacting agents' behaviors, where each agent's behavior affects and is affected by the aggregate behavior of the entire population by the concept of the mean-field term. This framework offers a powerful tool for understanding complex systems with numerous interacting components. The combination of ML and mean field game theory opens up new avenues for research and application development. By leveraging the principles of mean-field game theory, ML algorithms can be extended to handle large-scale problems with multiple interacting agents, enabling the modeling and optimization of complex systems in a more realistic and accurate manner. The integration of these two disciplines allows us to capture the inherent interdependencies and strategic interactions among the agents, leading to improved decision-making, resource allocation, and system control. This synergistic approach enhances the scalability, efficiency, and robustness of ML models, making them more adept at handling large-scale, distributed, and dynamic environments.

1.3 Challenges and Motivations

As powerful and transformative as ML is, it is not without its challenges and hurdles. The obstacles in the path of ML range from dealing with high-dimensionality, handling complex interactions in multi-agent systems, to privacy and security concerns in distributed learning. These challenges serve as the catalyst for the integration of MFG theory into ML.

One of the primary challenges in machine learning is the high computational complexity, especially in high-dimensional input data space. With the rapid development of technology and the widespread adoption of IoT devices, a vast amount of data is being generated at an unprecedented rate. These IoT devices, along with other sources such as social media platforms, sensors, and online services, contribute to an ever-expanding digital universe. The sheer volume of data being produced is staggering, creating new opportunities and challenges for ML systems. Moreover, the dimensions of the data are also increasing significantly, e.g., high-resolution images, videos, and other forms of rich media. This wealth of high-dimensional data requires advanced techniques and tools to address the significant computational complexity that it brings. Techniques like dimensionality reduction, feature selection, and efficient algorithms are critical but often fall short when the data is exceedingly large and complex.

Another significant challenge arises in large-scale multi-agent ML systems, particularly in the context of multi-agent reinforcement learning (RL). These systems consist of multiple agents, each operating independently, resulting in complex and unpredictable dynamics. Effectively modeling this intricate interaction space, coordinating strategies among agents, and promoting cooperation or competition is a substantial challenge that directly impacts the efficiency and effectiveness of the learning process [66, 140]. The difficulty intensifies when dealing with a large number of agents, exacerbating the complexity of the situation. The challenge is to manage these interactions in a way that ensures stability and convergence, especially when agents have conflicting objectives or need to collaborate to achieve a common goal [20, 61]. Traditional RL algorithms often struggle with the scalability and complexity of multi-agent environments, leading to suboptimal performance and slow convergence.

On the other hand, the decentralization characteristic of federated learning (FL) brings several challenges. Here, diverse heterogeneous devices with local datasets contribute to a global model without sharing raw data. While this approach aids privacy preservation, it simultaneously grapples with managing interactions between heterogeneous devices with possibly uneven computational and communication capabilities, data distributions, and physical locations. This heterogeneity complicates the process of model training, making it more difficult to achieve a high-performing global model. Challenges in FL also include dealing with non-IID (independent and identically distributed) data across clients, ensuring robust aggregation methods, and managing communication overhead effectively. Furthermore, the varying reliability and availability of devices add another layer of complexity, as the FL system must handle dropouts and asynchronous updates [102, 126, 145].

These challenges motivate us to look beyond traditional machine learning approaches and explore more powerful mathematical tools: MFG theory. MFG theory, with its ability to handle high-dimensional systems efficiently, offers a potential solution to the computational complexity challenge. It exploits the collective behavior of a large population of agents by approximating the group behavior of all agents through the concept of the mean-field approximation. In a machine learning system, the agents may refer to the training samples, the training batches, the agents in reinforcement learning, or the local clients in federated learning. From a game perspective, ML training is a process where a large number of agents interact with others. When the number of agents is huge, the ML system's structural complexity or computational complexity increases significantly.

In supervised learning, MFG provides a new perspective to study the interaction of the input data. By leveraging MFG control, the input data can be augmented and pre-processed based on the whole data distribution, improving the model's performance and robustness. This approach can lead to more accurate and generalized models that can better handle unseen data. Additionally, MFG can assist in optimizing the feature selection process by understanding the influence of each feature within the collective dataset.

In a multi-agent reinforcement learning system, MFG provides a structured method to model agent interactions. By considering the collective behavior of all agents, it offers a mathematical approximation for predicting and managing complex dynamics, thus simplifying system complexity and enhancing the learning process. MFG can help in developing strategies that are not only optimal but also stable and robust in highly dynamic environments. This is particularly useful in applications such as autonomous driving, where multiple vehicles must interact and coordinate their actions in real-time.

For federated learning, viewing each local client as a player in a game can bring a fresh perspective to handle heterogeneity. An MFG formulation can help balance the local and global objectives such as resource limitations and data contributions, taking into account the specific characteristics of each device and enabling more effective collaboration. MFG can also aid in designing aggregation methods that are more resilient to the variability in data quality and quantity across different clients, thereby enhancing the overall performance and robustness of the FL system.

In essence, the impetus to incorporate MFG into machine learning emerges from our endeavor to surmount these hurdles and enrich the capabilities of ML systems. The ensuing chapters of this book aim to illustrate the fusion of MFG and ML, examining the challenges at hand, and elucidating how MFG could provide innovative solutions for traditional ML. By integrating MFG with ML, we aim to create more scalable, efficient, and robust models capable of addressing complex, large-scale problems in various domains. This integration promises to push the boundaries of what is achievable in machine learning, leading to advancements that can tackle some of the most pressing challenges in today's data-driven world.

1.4 Chapter Organization

In this book, we delve into the exciting realm of mean-field game applications in machine learning. We explore how the principles of mean-field game theory can be harnessed to tackle real-world challenges and enhance the capabilities of machine learning algorithms. Through a combination of theoretical foundations, practical examples, and cutting-edge research, we aim to provide a comprehensive understanding of the intersection between mean field games and machine learning, offering insights and techniques that can be applied to a wide range of domains.

The organization of this book is as follows: Chap. 1 establishes the foundation by offering a comprehensive introduction to both MFG theory and the fundamentals of ML, along with the necessary prerequisites for understanding the material. Following this, Chap. 2 explores recent advancements in both fields, highlighting the challenges and motivations for integrating these domains. Subsequent chapters, from Chaps. 3 to 5, are dedicated to some specific applications and methodologies. These include the use of Generative Adversarial Networks (GANs) for opinion evolution in social networks, data augmentation techniques utilizing MFG, and the development of deep reinforcement learning frameworks guided by MFG principles. Moreover, Chaps. 6 and 7 cover innovative approaches such as incentive mechanism design and client selection in hierarchical federated learning using MFGs, and Chap. 8 discusses the application of MFG on evolutionary neural architecture search.

Whether you are a researcher, a practitioner, or simply curious about the latest advancements in machine learning and its applications, this book will serve as a valuable resource to explore the fascinating world of mean-field games and their integration with machine learning. By the end of this journey, you will have a solid foundation to apply mean-field game techniques in your own projects and be equipped to tackle complex problems that involve large-scale interactions and decision-making within a dynamic environment.

Let us now embark on this exciting journey into the realm of mean-field game applications in machine learning and unlock the potential for groundbreaking advancements in understanding and controlling large-scale distributed systems.

Chapter 2
Mean Field Game and Machine Learning Basis

In this chapter, we will explore the foundational concepts of MFGs and how these concepts intersect with the field of machine learning. Mean-field games provide a powerful framework for modeling and analyzing systems with a large number of interacting agents, making them particularly useful for applications in economics, finance, and various engineering domains. By understanding the principles of MFG, we can better appreciate their potential to enhance machine learning algorithms.

Mean-field theory originated in statistical physics as a method to analyze large-scale systems by averaging the effects of all individual components. This approach reduces the complexity inherent in dealing with numerous interactions, making it more tractable to derive meaningful insights and solutions. In recent years, the principles of mean-field theory have been adapted to study strategic interactions in large populations of agents, leading to the development of mean-field games. On the other hand, ML, a subset of artificial intelligence, focuses on building systems capable of learning from data and making predictions or decisions. As ML continues to evolve, the integration of mean-field theory offers promising enhancements, particularly in dealing with complex, multi-agent environments. The synergy between MFG and ML can lead to innovative solutions that address some of the most challenging problems in data science and artificial intelligence.

In this chapter, we will first provide an overview of mean-field games, introducing key concepts and mathematical formulations in Sect. 2.1. We will then delve into the basics of machine learning, outlining the primary types of learning algorithms and their applications in Sect. 2.2. In Sect. 2.4, we explore the intersection of mean-field games and machine learning, highlighting how these fields can complement each other to develop more robust and scalable models. Finally, we summarize this Chapter in Sect. 2.5.

By the end of this chapter, readers should have a solid understanding of the basics of mean-field games and their relevance to machine learning. This foundation will be crucial as we progress to more advanced topics and applications in subsequent chapters.

Y. Kang et al., *Mean Field Guided Machine Learning*, Wireless Networks,
https://doi.org/10.1007/978-3-031-91859-9_2

2.1 Mean Field Game Basis

Mean-field games (MFGs) were first proposed in [117] as a way to address the optimal control problem involving a large number of *indistinguishable* and *rational players*. "Indistinguishable" means that players share a common structure of a control model with *heterogeneous* states, while "rational" means that players act optimally to maximize their utility. Instead of assessing the control and status of each individual player, the generic player only responds to the collective behavior of all other players, referred to as the "mean field." In the context of a specific optimal control problem, the "mean field" usually refers to the discrete or continuous probability distribution of the state variable. The overall framework of MFG is shown in Fig. 2.1. By solving a system of partial differential equations (PDEs), named the Hamilton-Jacobi-Bellman (HJB) equations, players can determine their optimal control based on the mean field and their personal information. Upon making their decisions, the mean field can transform into a new distribution by solving the Fokker-Planck-Kolmogorov (FPK) equation [37]. By iteratively updating the HJB and FPK equations, the population can reach a Nash Equilibrium (NE), thus providing the optimal control for all the players.

In this subchapter, we first introduce the MFG on graphs with a discrete FPK as a well-defined gradient flow. Then, MFGs in continuous time and space are briefly introduced. After that, we will provide an introduction to some basics of machine learning techniques.

2.1.1 Discrete Mean Field Game on Graph

A discrete MFG is a game played by a large number of players in the player set $\mathcal{M} = \{1, 2, \ldots, M\}$. Each player in the population selects a strategy from the discrete

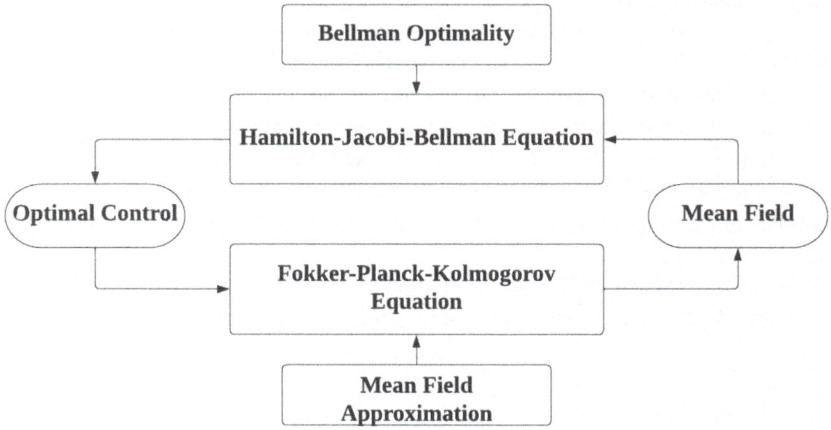

Fig. 2.1 Overall framework of MFG

strategy set $\mathcal{S} = \{1, 2, \ldots, S\}$, where S is the size of each player's strategy space. The payoff function of the players choosing strategy s is denoted by $F_s : \mathcal{P}(\mathcal{S}) \to \mathbb{R}$, where $\mathcal{P}(\mathcal{S})$ is the aggregated strategy of the population, which is given by

$$\mathcal{P}(\mathbb{S}) = \{(\rho_s)_{s=1}^{S} : \sum_{s=1}^{S} \rho_s = 1, \rho_s \geq 0, s \in \mathbb{S}\}, \tag{2.1}$$

where ρ_s is the probability of the players in $\mathcal{M}$ selecting the strategy s.

Let $\mathcal{P}_s$ be the set of all possible strategies for player s, and $P = (P_s, P_{-s})$ be a strategy profile, a set consisting of player s's strategy P_s and all the other players' strategies P_{-s}. Based on [123], the game reaches an NE $P^* = (P_s^*, P_{-s}^*)$ if:

$$F_s(P_s^*, P_{-s}^*) \geq F_s(P_s, P_{-s}^*), \text{for all } P_s \in \mathcal{P}_s. \tag{2.2}$$

Based on the above definitions of discrete MFGs, we then construct the strategy graph $\mathcal{G} = (\mathbb{S}, \mathbb{E})$:

- **Nodes and Edges:** Nodes of this graph are pure strategies from the discrete strategy set $\mathbb{S} = \{1, 2, \cdots, S\}$. Edges are connections between nodes. Node $i \in \mathbb{S}$ and node $j \in \mathbb{S}$ are able to form an edge $(i, j) \in \mathbb{E}$ if players can directly switch from strategy i to strategy j.
- **Neighborhood:** The neighborhood of node i is the set of all nodes which have a direct connection to node i. It's defined as

$$N(i) = \{j \in \mathbb{S} : (i, j) \in \mathbb{E}\}.$$

- **State Space:** The state space of the strategy graph consists of all available mixed strategies of the players, i.e., $\mathcal{P}(s)$ defined in (2.1)

Given the discrete strategy set $\mathbb{S}$ and population state space $\mathcal{P}(\mathbb{S})$, we can introduce Theorem 2.1 as follows:

Theorem 2.1 *Suppose that a population game has strategy graph $\mathcal{G} = (\mathbb{S}, \mathbb{E})$ and the constant $\beta \geq 0$, the utility function $F : \mathcal{P}(\mathbb{S}) \to \mathbb{R}^n$ are continuous. Then for any initial condition $\rho^0 \in \mathcal{P}_o(\mathbb{S})$, the Fokker-Planck-Kolomogorov equation*

$$\begin{aligned}\frac{\mathrm{d}\rho_i}{\mathrm{d}t} = &\sum_{j \in N(i)} \frac{1}{d_j} \rho_j [F_i(\rho) - F_j(\rho) + \beta(\log \rho_j - \log \rho_i)]^+ \\ &- \sum_{j \in N(i)} \frac{1}{d_i} \rho_i [F_j(\rho) - F_i(\rho) + \beta(\log \rho_i - \log \rho_j)]^+\end{aligned} \tag{2.3}$$

is a well-defined gradient flow in $\mathcal{P}_o(\mathbb{S})$. β represents the strength of uncertainty, $[\cdot]^+ = max\{\cdot, 0\}$, $N(i)$ is the neighborhood of node i, and the degree of node j is denoted as $d_j = \sum_{i \in N(j)} 1$.

The discrete FPK in (2.3) is a pairwise comparison evolutionary dynamics [163], which describes the population evolution in the state space $\mathcal{P}(\mathbb{S})$, i.e., the evolution of the mean field. During each iteration, we will compare the potential of selecting strategy i and strategy j, which is measured by a utility function corresponding to the specific application. The selection probability that flows from current strategy i to candidate strategy j is proportional to the difference between their strategy payoffs (their potentials).

2.1.2 *Mean Field Game in Continuous Time and Space*

MFGs in continuous time and space typically consist of infinitely many rational asymptotically negligible agents, denoted as $\mathcal{A} = \{a_1, a_2, \ldots, a_k\}$. The dynamics of the agents is governed by a stochastic differential equation given as

$$dx^k(t) = f\left(x^k(t), u^k(t)\right) dt + \sigma dW(t), \tag{2.4}$$

where $x^k(t)$ is the state of the agent a_k, $u^k(t)$ is the control input of agent a_k, and $w(t)$ is the standard Brownian motion with intensity σ. We assume that σ is the same for all agents in $\mathcal{A}$.

The main concept of MFGs is the mean-field term, which is defined as the distribution over the agents' state space of strategy space. When the number of agents is large, their locations and actions can be approximated as their distribution, that is

$$\rho(t, x) = \lim_{K \to \infty} \frac{1}{K} \sum_{k=1}^{K} \delta_{x^k(t)=x}. \tag{2.5}$$

Note that by using the mean-field term $\rho(t, x)$, we can measure the states and behaviors of all agents. In this way, we can restate the game model as an interaction between agents with the mass of others.

The goal of a reference agent is to find a control $u^k(t)$ which minimizes the cost function J_k shown as follow.

$$J_k = \mathbb{E}\left[\int_0^T L(x(t), u(t), \rho(t))dt + G\left(x(T), \rho(T)\right)\right], \tag{2.6}$$

where $\rho(t)$ is the distribution of agents, i.e., the mean-field term, and $\rho(T)$ is the final distribution of agents. $L(x(t), u(t), \rho(t))$ and $G(x(T), \rho(T))$ are called the running cost function and the terminal cost function, respectively.

The optimal strategy $u^*(t)$ can be derived by solving the following partial differential equations (PDEs).

$$-\partial_t\phi - \frac{\sigma^2}{2}\Delta\phi + H(x, \rho, \nabla\phi) = 0, \tag{2.7}$$

$$\partial_t\rho - \frac{\sigma^2}{2}\Delta\rho - \nabla\cdot\left(\rho\nabla_p H\left(x, \rho, \nabla\phi\right)\right) = 0, \tag{2.8}$$

$$\rho(0, x) = \rho_0(x),\ \phi(T, x) = G\left(X(T), \rho(T, x)\right), \tag{2.9}$$

where $\phi(t, x)$ is the value function defined as

$$\phi(t, x) = \inf_{u(t)}\left\{\mathbb{E}\left[\int_t^T f\left(u(t), X(t), \rho(t)\right)dt + G\left(X(T), \rho(T)\right)\right]\right\}. \tag{2.10}$$

$H(x, \rho, p)$ is the Hamiltonian defined as

$$H(x, \rho, p) = \max_u\left\{-f(x, u)p - L(x, u, \rho)\right\}. \tag{2.11}$$

Equation (2.7) is the HJB equation. It is a PDE that gives a necessary and sufficient condition for optimality of a general MFG problem, and is derived by the principle of optimality in dynamic programming. By solving the HJB equation, the optimal value function $\phi^*(t, x)$ in (2.10) can be derived, and the optimal control $u^*(t)$ is then derived by taking the maximizer of the Hamitonian in (2.11). Equation (2.8) is the FPK equation. It is also a PDE that describes the time evolution of the probability density function of agents' states $x(t)$ under the influence of the control $u(t)$ and Brownian motion $W(t)$, and is derived by the agents dynamics in (2.4). Equation (2.9) is the boundary condition for ρ and ϕ. By solving the above PDEs, the optimal control $u^*(t)$ and state $x^*(t)$ can be derived.

A classical MFG example in continuous time and space is a variational problem akin to the Benamou-Brenier formulation for the Optimal Transport (OT) problem [133]

$$\begin{aligned}&\inf_{\rho,u}\int_0^T\left\{\int_\Omega \rho(x,t)L(x,u(x,t))\,dx + \mathcal{F}(\rho(x,t))\right\}dt + \mathcal{G}(\rho(x,T)),\\ &\text{s.t. } \frac{\partial\rho}{\partial t} + \nabla\cdot(\rho u) - \eta\Delta\rho = 0,\ \rho(0,x) = \rho_0,\end{aligned} \tag{2.12}$$

where $x \in \mathbb{R}^n$ is the state variable; $\rho : [0, T] \times \mathbb{R}^n \to \mathbb{R}$ is the probability density function of x; $u : \mathbb{R}^n \times [0, T] \to \mathbb{R}^n$ is the control; $L : \mathbb{R}^n \times \mathbb{R}^n \to \mathbb{R}$ is the running cost, which is also the Lagrangian function related to the Legendre transform of the Hamiltonian H; $\mathcal{F} : \mathbb{R}^n \times \mathcal{P}(\mathbb{R}^n) \to \mathbb{R}$ denotes the interaction between agents and the population; $\mathcal{G} : \mathbb{R}^n \times \mathcal{P}(\mathbb{R}^n) \to \mathbb{R}$ is the terminal cost; and ρ_0 is the initial condition of ρ. Following the Bellman's idea of dynamic programming and the principle of optimality [122], we define the value function as

$$\Phi(x,t) = \inf_{u[t,T]} \left\{ \int_t^T \left\{ \int_\Omega \rho(x,t) L(x, u(x,t))\, dx + \mathcal{F}(\rho(x,t)) \right\} dt + \mathcal{G}(\rho(x,T)) \right\}, \tag{2.13}$$

where $u[t, T]$ is the control from any time slot t to the terminal time T and $\Phi(x, t)$ is the optimal cost-to-go from (x, t). Solving (2.12) is equivalent to solve the system of partial differential equations (PDEs) as follows:

$$\begin{aligned} & -\partial_t \Phi - \eta \Delta \Phi + H(x, \nabla \Phi) = \mathcal{F}(x, \rho), \quad \text{(HJB)} \\ & \partial_t \rho - \eta \Delta \rho - \nabla \cdot (\rho \nabla_p H(x, \nabla \Phi)) = 0, \quad \text{(FPK)} \\ & \rho(0, x) = \rho_0, \ \Phi(x, T) = \mathcal{G}(x, \rho^T). \end{aligned} \tag{2.14}$$

Any pair of the solution (ρ, Φ) to the original problem (2.12) should satisfy the HJB and FPK equations in (2.14). The convergence of the FPK and HJB residual errors can reveal the convergence of the MFG.

2.2 Machine Learning Basis

ML is a transformative technology that has revolutionized various fields by enabling computers to learn from data and make decisions or predictions without being explicitly programmed. At its core, ML involves developing algorithms that can identify patterns within data, allowing systems to improve their performance over time through experience. This ability to learn and adapt makes ML a powerful tool for solving complex problems across numerous domains, including healthcare, finance, autonomous systems, and more.

In this subchapter, we will delve into the basics of these key areas in machine learning, exploring their principles, algorithms, and applications. Understanding these concepts is essential for leveraging the full potential of machine learning to tackle real-world challenges and harness the power of data-driven decision-making.

2.3 Supervised Learning

Supervised learning is a fundamental approach in machine learning where the goal is to learn a mapping from inputs to outputs based on example input-output pairs. This technique has been widely used in various applications, ranging from image recognition and natural language processing to medical diagnostics and financial forecasting [47].

In supervised learning, we have a dataset consisting of input-output pairs $(\mathbf{x}_i, y_i)$ where $\mathbf{x}_i$ represents the input features and y_i represents the corresponding output or label. The objective is to learn a function f that maps inputs to outputs, such that $f(\mathbf{x}_i) \approx y_i$. This can be formalized as finding the function f that minimizes a loss function over the training data.

Let $\mathcal{D} = \{(\mathbf{x}_i, y_i)\}_{i=1}^n$ be the training dataset, where $\mathbf{x}_i \in \mathbb{R}^d$ and $y_i \in \mathbb{R}$ for regression tasks or $y_i \in \{1, 2, \dots, K\}$ for classification tasks. The goal is to learn a function $f : \mathbb{R}^d \rightarrow \mathbb{R}$ (for regression) or $f : \mathbb{R}^d \rightarrow \{1, 2, \dots, K\}$ (for classification).

The learning problem can be formulated as the following optimization problem:

$$\min_{f \in \mathcal{F}} \frac{1}{n} \sum_{i=1}^{n} \mathcal{L}(f(\mathbf{x}_i), y_i), \tag{2.15}$$

where $\mathcal{F}$ is the hypothesis space of possible functions and $\mathcal{L}$ is the loss function that measures the discrepancy between the predicted output $f(\mathbf{x}_i)$ and the true output y_i.

The supervised learning process starts with collecting a labeled dataset of input-output pairs. An appropriate algorithm is chosen based on the problem type. The model is then trained on the training data by optimizing parameters to minimize prediction errors. After training, the model is evaluated on the testing set using metrics like accuracy or mean squared error. Hyperparameters are fine-tuned to improve performance. Once validated, the model is used to make predictions on new data and is deployed to a production environment. Continuous monitoring and maintenance ensure the model remains accurate and relevant.

2.3.1 Supervised Learning Algorithms

Supervised learning is a fundamental machine learning approach where a model is trained on labeled data to learn the mapping between inputs and outputs. This technique is widely used across various domains for tasks such as classification and regression. In this section, we will explore several key supervised learning algorithms, including Linear Regression, Logistic Regression, Support Vector Machines (SVM), and various types of Neural Networks, including Convolutional Neural Networks (CNNs), Residual Networks (ResNets), and Transformers.

2.3.1.1 Linear Regression

Linear regression is used for predicting a continuous target variable. It assumes a linear relationship between the input features and the target variable. The model can be represented as:

$$y = \mathbf{w}^\top \mathbf{x} + b, \tag{2.16}$$

where y is the predicted output, $\mathbf{w}$ is the weight vector, $\mathbf{x}$ is the input feature vector, and b is the bias term. The objective of linear regression is to find the parameters $(\mathbf{w}, b)$ that minimize the mean squared error (MSE) between the predicted outputs and the actual target values:

$$\min_{\mathbf{w},b} \frac{1}{n} \sum_{i=1}^{n} (y_i - \mathbf{w}^\top \mathbf{x}_i - b)^2, \tag{2.17}$$

where n is the number of training samples, y_i is the actual target value, and $\mathbf{x}_i$ is the feature vector for the i-th sample.

2.3.1.2 Logistic Regression

Logistic regression is used for binary classification tasks. It models the probability that a given input belongs to a particular class. The model predicts the probability using the sigmoid function:

$$P(y = 1|\mathbf{x}) = \sigma(\mathbf{w}^\top \mathbf{x} + b), \tag{2.18}$$

where $\sigma(z) = \frac{1}{1+e^{-z}}$ is the sigmoid function, $\mathbf{w}$ is the weight vector, $\mathbf{x}$ is the input feature vector, and b is the bias term. The objective is to maximize the likelihood of the observed data, which is equivalent to minimizing the binary cross-entropy loss:

$$\min_{\mathbf{w},b} -\frac{1}{n} \sum_{i=1}^{n} \left[y_i \log \sigma(\mathbf{w}^\top \mathbf{x}_i + b) + (1 - y_i) \log(1 - \sigma(\mathbf{w}^\top \mathbf{x}_i + b)) \right], \tag{2.19}$$

where y_i is the actual binary label for the i-th sample.

2.3.1.3 Support Vector Machines (SVM)

Support Vector Machines are powerful models used for classification and regression tasks. The goal of SVM is to find the optimal hyperplane that separates data points of different classes with the maximum margin. For linearly separable data, the decision boundary is defined as:

$$\mathbf{w}^\top \mathbf{x} + b = 0, \tag{2.20}$$

where $\mathbf{w}$ is the weight vector and b is the bias. The margin is the distance between the hyperplane and the nearest data points from each class, known as support vectors. The objective is to maximize this margin, which can be formulated as:

$$\min_{\mathbf{w},b} \frac{1}{2}\|\mathbf{w}\|^2, \tag{2.21}$$

subject to the constraint that all data points are correctly classified, i.e., $y_i(\mathbf{w}^\top \mathbf{x}_i + b) \geq 1$ for all i.

For non-linearly separable data, SVM can be extended using kernel functions that map the input features into a higher-dimensional space where a linear separator can be found. Common kernels include the polynomial kernel and the radial basis function (RBF) kernel.

2.3.1.4 Neural Networks

Neural networks consist of layers of interconnected neurons, where each neuron performs a weighted sum of its inputs followed by a non-linear activation function. They are capable of capturing complex non-linear relationships and are used for a wide range of tasks, from image recognition to natural language processing.

A simple neural network consists of an input layer, one or more hidden layers, and an output layer. The output of a neuron i in layer l can be computed as:

$$a_i^{(l)} = \sigma\left(\sum_j w_{ij}^{(l-1)} a_j^{(l-1)} + b_i^{(l)}\right), \tag{2.22}$$

where σ is the activation function (e.g., ReLU, sigmoid), $w_{ij}^{(l-1)}$ is the weight connecting neuron j in layer $l-1$ to neuron i in layer l, $a_j^{(l-1)}$ is the activation of neuron j in layer $l-1$, and $b_i^{(l)}$ is the bias of neuron i in layer l. Many different types of neural networks have been proposed in existing works. We give a brief introduction to some well-adopted types below.

Convolutional Neural Networks (CNNs) are a specialized type of neural network designed for processing structured grid data, such as images. CNNs use convolutional layers that apply a set of filters (kernels) to the input image to extract features such as edges, textures, and patterns.

A convolution operation on an input $\mathbf{x}$ with a filter $\mathbf{w}$ produces an output feature map $\mathbf{y}$:

$$y_{ij} = (\mathbf{x} * \mathbf{w})_{ij} = \sum_m \sum_n x_{i+m,j+n} w_{mn}, \tag{2.23}$$

where $*$ denotes the convolution operation. CNNs also use pooling layers to reduce the spatial dimensions of the feature maps, which helps to make the model invariant to small translations of the input.

Residual Networks (ResNets) are a type of deep neural network designed to address the vanishing gradient problem, which makes training very deep networks difficult. ResNets introduce shortcut connections (or residual connections) that bypass one or more layers.

The main idea is to learn residual functions with reference to the layer inputs, instead of learning unreferenced functions. A building block of a ResNet can be represented as:

$$\mathbf{y} = \mathcal{F}(\mathbf{x}, \{W_i\}) + \mathbf{x}, \tag{2.24}$$

where $\mathbf{x}$ is the input, $\mathcal{F}$ is the residual mapping to be learned, and $\{W_i\}$ are the weights of the layers in the residual block.

Transformers are a type of neural network architecture primarily used in natural language processing (NLP) tasks. They rely on a mechanism called self-attention to weigh the influence of different words in a sentence when encoding and decoding sequences.

The self-attention mechanism calculates a set of attention scores that determine how much each word should focus on every other word in the sequence. The core components of a transformer include the encoder and decoder, each consisting of multiple layers of self-attention and feed-forward neural networks.

The self-attention mechanism can be represented as:

$$\text{Attention}(Q, K, V) = \text{softmax}\left(\frac{QK^\top}{\sqrt{d_k}}\right)V, \tag{2.25}$$

where Q (queries), K (keys), and V (values) are linear projections of the input, and d_k is the dimension of the keys.

These supervised learning algorithms are foundational to many practical applications. Each algorithm has its strengths and is suited to different types of problems. Understanding their principles, strengths, and limitations allows practitioners to choose and implement the most appropriate method for their specific use case.

2.3.2 Reinforcement Learning

Reinforcement learning (RL) is a branch of machine learning developed to enable agents to learn optimal behaviors through interactions with an environment. Unlike supervised learning, where the model learns from labeled examples, RL agents learn by receiving feedback in the form of rewards or punishments based on their actions. This trial-and-error approach allows the agent to develop strategies that maximize cumulative rewards over time. RL has been successfully applied in various domains such as robotics [112], game playing (e.g., AlphaGo [172]), autonomous driving [111], and personalized recommendations [216]. Its importance lies in its ability to solve complex sequential decision-making problems where the optimal solution is not immediately apparent and must be discovered through exploration.

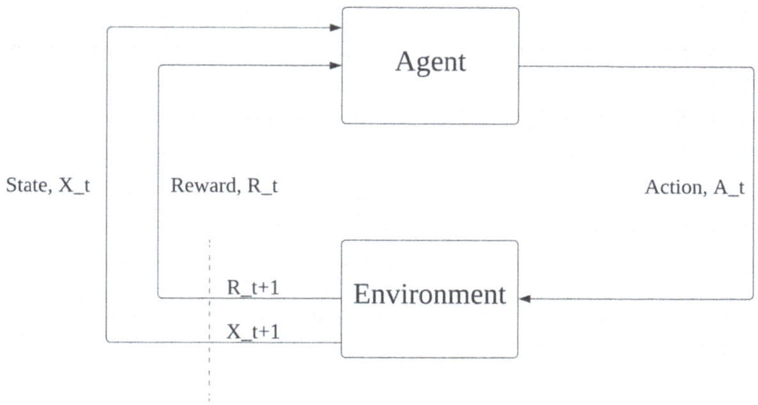

Fig. 2.2 The general framework of Markov decision process

2.3.2.1 Markov Decision Processes

The general reinforcement learning process is established based on Markov decision process (MDP) as shown in Fig. 2.2. *Agents* take *actions* and reach a new state of the *environment*, which will give the agent a *reward* as feedback [176].

Mathematically, an MDP is defined by the tuple (S, A, P, R, γ):

- S: A set of states representing the environment.
- A: A set of actions available to the agent.
- $P(s'|s, a)$: The transition probability of moving from state s to state s' after taking action a.
- $R(s, a)$: The reward received after taking action a in state s.
- γ: The discount factor, which balances immediate and future rewards ($0 \leq \gamma \leq 1$).

In an MDP, the agent interacts with the environment in discrete time steps. At each time step t, the agent observes the current state s_t, chooses an action a_t, and receives a reward $R(s_t, a_t)$. The environment then transitions to a new state s_{t+1} according to the transition probability $P(s_{t+1}|s_t, a_t)$.

2.3.2.2 Mathematical Formulation

The objective in RL is to learn a policy $\pi(a|s)$, which defines the probability of taking action a in state s to maximize the expected cumulative reward, also known as the return. The return G_t at time t is given by:

$$G_t = \sum_{k=0}^{\infty} \gamma^k R_{t+k+1}, \tag{2.26}$$

where R_{t+k+1} is the reward received $k+1$ steps into the future.

Two key functions in RL are the state-value function $V^{\pi}(s)$ and the action-value function $Q^{\pi}(s,a)$:

The state-value function $V^{\pi}(s)$ represents the expected return starting from state s and following policy π:

$$V^{\pi}(s) = \mathbb{E}^{\pi}\left[G_t \mid S_t = s\right] = \mathbb{E}^{\pi}\left[\sum_{k=0}^{\infty} \gamma^k R_{t+k+1} \mid S_t = s\right]. \tag{2.27}$$

The action-value function $Q^{\pi}(s,a)$ represents the expected return starting from state s, taking action a, and then following policy π :

$$Q^{\pi}(s,a) = \mathbb{E}^{\pi}\left[G_t \mid S_t = s, A_t = a\right] = \mathbb{E}^{\pi}\left[\sum_{k=0}^{\infty} \gamma^k R_{t+k+1} \mid S_t = s, A_t = a\right]. \tag{2.28}$$

The goal is to find the optimal policy π^* that maximizes the expected return from any state s:

$$\pi^* = \arg\max_{\pi} V^{\pi}(s) \text{ for all } s \in S. \tag{2.29}$$

Common algorithms for finding the optimal policy include Q-Learning and Policy Gradient methods. Q-Learning updates the action-value function based on the Bellman equation:

$$Q(s_t, a_t) \leftarrow Q(s_t, a_t) + \alpha\left[R_{t+1} + \gamma \max_{a'} Q(s_{t+1}, a') - Q(s_t, a_t)\right], \tag{2.30}$$

where α is the learning rate.

Policy Gradient methods directly optimize the policy by adjusting the policy parameters θ to maximize the expected return:

$$\nabla_{\theta} J(\theta) = \mathbb{E}_{\pi}\left[\nabla_{\theta} \log \pi_{\theta}(a|s) Q^{\pi}(s,a)\right], \tag{2.31}$$

where $J(\theta)$ is the objective function representing the expected return.

2.3.3 Generative Adversarial Networks

Generative Adversarial Networks (GANs), first proposed by Goodfellow et al. [80] in 2004, are generative models designed to approximate an unknown, usually high-dimensional probability distribution P_r by generating a parameterized probability distribution P_{θ}. The overall framework is shown in Fig. 2.3. GANs leverage the

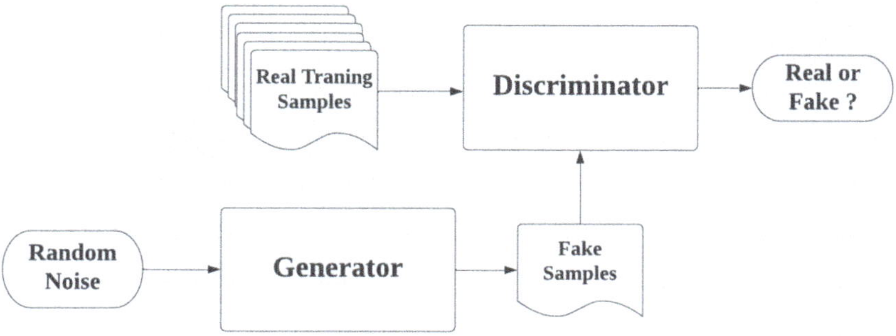

Fig. 2.3 Overall framework of GAN

benefits of piecewise linear units and address the difficulties of approximating intractable probabilistic computations through a framework consisting of two adversarial Neural Networks (NNs): a generator network G_θ and a discriminator D_ω, where θ and ω are parameters. The generator is used to generate fake samples from a given latent space $\mathcal{Z}$, with a similar distribution to P_r, i.e., $G_\theta : \mathcal{Z} \rightarrow \mathcal{X}$. The discriminator is trained to distinguish the generated fake samples from the real samples, estimating the probability that an input sample is real, i.e., $D_\omega : \mathcal{X} \rightarrow [0, 1]$. During training, the generator seeks to maximize the probability of D_ω making mistakes, while the discriminator attempts to estimate the probability that a sample comes from the real distribution P_r instead of the generator. The training process terminates when the discriminator can no longer distinguish the fake samples from the real ones, i.e., when the function D_ω equals $\frac{1}{2}$. Mathematically, the goal of the original GAN is to find an equilibrium to the two-player minimax game

$$\inf_{G_\theta} \sup_{D_\omega} \mathbb{E}_{x \sim P_r}[\log D_\omega(x)] + E_{z \sim \mathcal{Z}}[\log(1 - D_\omega(G_\theta(z)))]. \tag{2.32}$$

2.3.4 *Federated Learning*

Federated Learning (FL) is a decentralized approach to machine learning that enables the training of models across multiple devices or servers holding local data samples, without exchanging them [145]. This method was developed to address growing concerns about data privacy, security, and data ownership. In traditional centralized machine learning, data is collected and stored in a central server, which can pose significant risks related to data breaches and misuse. Federated Learning mitigates these risks by keeping the data localized and only sharing model updates, such as gradients or weights, thereby preserving user privacy and complying with data protection regulations.

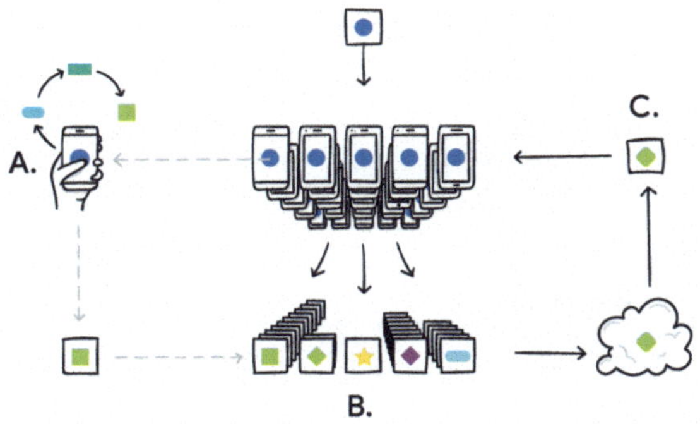

Fig. 2.4 Overall framework of federated learning [18]

The concept of Federated Learning was introduced by Google in 2016 to improve the text prediction performance of machine learning models on mobile devices without compromising user data privacy, as shown in Fig. 2.4. Since then, FL has been applied in various domains such as healthcare [170], finance [200], IoT [19], and autonomous vehicles [158], where data is distributed across multiple sources and privacy concerns are paramount. FL represents a significant advancement in machine learning, enabling the development of robust and privacy-preserving models across distributed data sources. By leveraging the principles demonstrated in the FL algorithms, FL addresses the challenges of communication efficiency, data heterogeneity, and privacy, paving the way for broader applications in various domains.

The Federated Learning Process Federated Learning involves training a global model collaboratively across multiple clients (e.g., mobile devices) that each hold local data. A common approach to implementing FL is through the Federated Averaging (FedAvg) algorithm, which provides a clear example of how the process works:

1. *Local Training:* Each client trains a local model on its own data for a set number of epochs or iterations. This training is done independently and in parallel on each device.
2. *Model Update*: After local training, each client computes model updates, such as gradients or weights, and sends these updates to a central server. Importantly, the raw data never leaves the clients, preserving privacy.
3. *Aggregation*: The central server aggregates the received updates by averaging them to form a new global model. This aggregation step combines the learning from all clients without requiring access to their data.
4. *Distribution*: The updated global model is then distributed back to the clients. Each client updates its local model with the new global model parameters.
5. *Iteration*: This process of local training, model update, aggregation, and distribution is repeated for multiple communication rounds until the global model converges to a satisfactory performance level.

The update rule for the global model in FedAvg can be formalized as:

$$\mathbf{w}_{t+1} = \sum_{k=1}^{K} \frac{n_k}{n} \mathbf{w}_t^k, \tag{2.33}$$

Through the above processes, FL enhances data privacy and security, significantly reducing the risk of data breaches and misuse. This decentralized approach eliminates the need for centralized data storage, making it particularly valuable for handling sensitive and personal data. FL enables learning from diverse and distributed datasets, improving the generalization capabilities of models and supporting edge computing, which is crucial for real-time processing on distributed devices. Additionally, FL reduces communication costs and latency by minimizing the need for frequent data transfers between clients and servers, leading to more efficient and scalable machine learning solutions. These benefits make FL an ideal framework for applications in healthcare, finance, IoT, autonomous vehicles, and other domains where data privacy, security, and real-time processing are paramount.

2.4 Discuss of the Intersection of MFG with ML

The foundation of machine learning lies in several key types of learning: supervised learning, reinforcement learning, Generative Adversarial Networks (GANs), and federated learning. Each of these areas offers unique approaches to handling data and extracting insights, and they can be closely related to the concept of MFGs. In MFGs, a large number of agents interact in a way that the collective behavior of the agents influences the outcome for each individual agent. This concept can be mapped to various scenarios in ML:

- **Data Samples in Supervised Learning**: In supervised learning, each data sample can be seen as an agent. The learning algorithm can be viewed as a mean-field game where the goal is to optimize a collective objective function, such as minimizing prediction error. The mean field in this context represents the aggregated information from all data samples that guides the learning process.
- **Agents in Reinforcement Learning**: In reinforcement learning, agents learn to make decisions based on interactions with their environment. This can be likened to MFG, where each agent's strategy evolves in response to the aggregate behavior of all other agents. The mean field here can represent the overall state of the environment influenced by all agents' actions.
- **Generative Models in GANs**: In Generative Adversarial Networks (GANs), two neural networks–the generator and the discriminator–compete against each other. This adversarial process can be seen as a game where the generator and discriminator influence each other's strategies, similar to how agents in MFG respond to the collective behavior of others. The mean field in GANs can be

thought of as the distribution of generated data, which evolves as the generator improves to fool the discriminator.

- **Local Clients in Federated Learning**: In federated learning, local clients (e.g., mobile devices) train models on their own data and share model updates rather than raw data. This distributed learning approach can be seen as a mean-field game where each client acts as an agent. The mean field represents the aggregated model updates that are shared across all clients to improve the global model.

By understanding these parallels, we can leverage insights from mean-field game theory to develop more efficient and scalable machine learning algorithms. For instance, the iterative updating process in MFG, where agents continually adjust their strategies based on the mean field, mirrors how ML models are trained and refined over time through feedback and optimization.

2.5 Chapter Summary

In this chapter, we provided a comprehensive introduction to the foundational concepts of MFG and ML, focusing on their intersection and applications.

First, we explored the basics of MFG, which addresses optimal control problems involving a large number of indistinguishable and rational players. The concept of the "mean field" was introduced, representing the collective behavior of all players. We discussed the Hamilton-Jacobi-Bellman (HJB) equations for determining optimal control and the Fokker-Planck-Kolmogorov (FPK) equation for updating the mean field. Discrete and continuous MFGs were examined, highlighting their applications in various domains.

Next, we delved into the fundamentals of Machine Learning, covering key supervised learning algorithms such as Linear Regression, Logistic Regression, and Support Vector Machines (SVM). We also discussed advanced neural network architectures including Convolutional Neural Networks (CNNs), Residual Networks (ResNets), and Transformers, illustrating their capabilities in capturing complex patterns and relationships in data.

Reinforcement Learning (RL) was introduced, emphasizing its importance in learning optimal behaviors through interactions with an environment. The mathematical formulation of RL using Markov Decision Processes (MDPs) was presented, along with key algorithms like Q-Learning and Policy Gradient methods.

Then, we covered Federated Learning (FL), a decentralized approach to machine learning designed to address data privacy and security concerns. Using the Federated Averaging (FedAvg) algorithm as an example, we explained the FL process and its applications in domains such as healthcare, finance, IoT, and autonomous vehicles.

Finally, by integrating the concepts of MFG and ML, we highlighted the potential for developing robust and efficient models capable of addressing complex, real-world problems while ensuring data privacy and security. This chapter sets the stage for further exploration of advanced topics in machine learning and their applications in distributed and privacy-preserving settings.

Chapter 3
Opinion Evolution in Social Networks: Use Generative Adversarial Networks to Solve Mean Field Game

Belief and opinion evolution in social networks (SNs) can aid in understanding how people influence others' decisions through social relationships. As a large number of users are involved in SNs, the complexity of traditional optimization techniques is high as they deal with the interactions between users separately. Moreover, the state variable (opinion) is high-dimensional because a person usually has opinions about many different social issues. Incorporating classical opinion dynamics, we formulate the opinion evolution in SNs as a high-dimensional stochastic mean field game (MFG). Numerical methods for high-dimensional MFGs are practically non-existent because of the need for grid-based spatial discretization. Thus, we propose a machine-learning method to tractably solve high-dimensional stochastic MFGs. With this approach, solving MFGs can be regarded as a special case of training a generative adversarial network. In the simulation, we analyze the effect of random social issues and stubbornness on the opinion evolution. Moreover, with the Social Evolution data set, we show that the proposed algorithm can efficiently predict the diffusion of opinions in SNs.

3.1 Opinion Diffusion Model in Social Networks

The system model consists of the cost function and the opinion dynamics, which describe how SN users affect others' opinions through social relationships. We consider a SN with N users denoted by graph $\mathcal{G} = (\mathcal{V}, \mathcal{E})$, where $\mathcal{V}$ is the set of all SN users and $\mathcal{E} \subseteq \mathcal{V} \times \mathcal{V}$ is the set of all connections between users as shown in Fig. 3.1. We denote the continuous opinion space as Ω. The opinion of user i is denoted by $x_i \in \Omega$, where $i = 1, 2, \cdots, N$. The jth entry x_{ij}, $j = 1, 2, \cdots, n$ represents user i's opinion towards the jth social event.

Y. Kang et al., *Mean Field Guided Machine Learning*, Wireless Networks,
https://doi.org/10.1007/978-3-031-91859-9_3

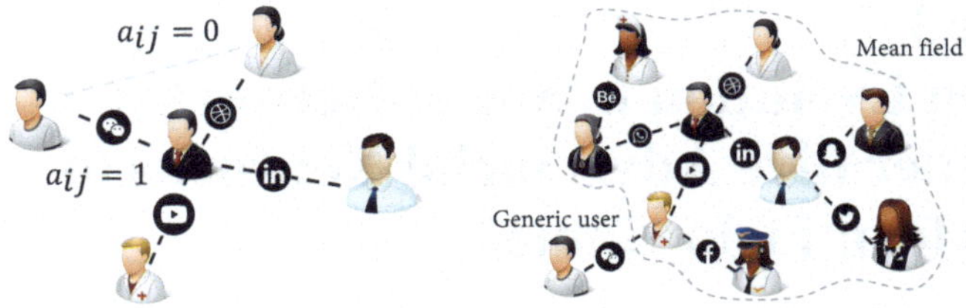

Fig. 3.1 An illustration of online social networks

3.1.1 Opinion Dynamics

Without considering stubbornness of SN users, the opinion x_i evolves according to opinion dynamics in the Hegselmann-Krause [87] model. The new opinion of user i at time $k + 1$ equals to the average opinion of user i's neighbors at time k, which is given by

$$x_i(k+1) = \frac{\sum_{j=1}^{N} \alpha_{ij} x_j(k)}{\sum_{j=1}^{N} \alpha_{ij}} + \beta_i u_i(k), \tag{3.1}$$

where β_i is a constant, u_i is the control implemented by user i in the view of the opinion x_i at time k. In the real life, u_i can refer to the actions that user i takes to determine her new opinion, such as browsing through a related magazine and consulting some related authorities. α_{ij} is a binary variable indicating the connection between users i and j. As shown in Fig. 3.1, $\alpha_{ij} = 0$ represents that user i and user j are not connected and they can not influence each other. $\alpha_{ij} = 1$ means that users i and j are connected with certain social relationships and they can influence each other. The physical meaning of (3.1) is that the new opinion of a non-stubborn user is determined by the average opinion of his/her neighbors and his/her control.

In the FJ model [68], the opinion of a stubborn user is determined by the average opinion of neighbors and the initial opinion. In our opinion model, we regard stubbornness as the fact that a stubborn user will always consider the current opinion, instead of the initial opinion, when he/she wants to change the opinion. The opinion dynamics are given by

$$x_i(k+1) = (1 - \lambda_i) \sum_{j=1}^{N} \omega_{ij} x_j(k) + \lambda_i x_i(k), \tag{3.2}$$

where λ_i denotes the strength of stubbornness, ω_{ij} refers to the interpersonal influence between users i and j, and $x_i(k)$ is the current opinion of user i. We integrate the concept of stubbornness to the HK model and obtain the opinion dynamics as follows

$$x_i(k+1) = (1-\lambda_i)\frac{\sum_{j=1}^{N} \alpha_{ij} x_j(k)}{\sum_{j=1}^{N} \alpha_{ij}} + \lambda_i x_i(k) + \beta_i u_i(k), \tag{3.3}$$

where we set the interpersonal influence to $\omega_{ij} = \alpha_{ij}/\sum_{j=1}^{N} \alpha_{ij}$ and add the personal effort of user i with strength β_i.

In continuous time, the corresponding opinion dynamics with stubborn SN users are given by

$$\frac{x_i(t+dt) - x_i(t)}{dt} = (1-\lambda_i)\frac{\sum_{j=1}^{N} \alpha_{ij} x_j(t)}{\sum_{j=1}^{N} \alpha_{ij}} + (\lambda_i - 1)x_i(t) + \beta_i u_i(t), \tag{3.4}$$

When λ_i (the stubbornness of user i) decreases to 0, (3.4) can represent the opinion dynamics of non-stubborn users. When the time interval $dt \to 0$, the opinion dynamics in (3.4) become

$$\begin{aligned} dx_i(t) &= \left((1-\lambda_i)\frac{\sum_{j=1}^{N} \alpha_{ij} x_j(t)}{\sum_{j=1}^{N} \alpha_{ij}} + (\lambda_i - 1)x_i(t) + \beta_i u_i(t) \right) dt \\ &= f_i(x_i(t), x_i^{\mathcal{D}_i}(t), u_i(t))dt, \end{aligned} \tag{3.5}$$

where $\mathcal{D}_i$ is the neighborhood of user i and $x_i^{\mathcal{D}_i}(t)$ represents the opinions of user i's neighbors.

3.1.2 Cost Function of a SN User

The cost functional of user i is given by

$$\mathcal{J}_i(x_i, u_i) = \int_0^T L_i(x_i, u_i)dt + \mathcal{G}_i(x_i^T), \tag{3.6}$$

where L_i is the running cost and G_i is the terminal cost. The running cost L_i is given by

$$L_i(x_i, u_i) = c_1 ||u_i(t)||_2^2, \tag{3.7}$$

where c_1 is a constant. The running cost L_i denotes the control effort implemented by user i in the view of his/her own opinion and other users' opinions at time t. The terminal cost $\mathcal{G}_i$ is given by

$$\mathcal{G}_i((x_i)^T) = c_2 ||x_i^T - x_T||_2, \tag{3.8}$$

where c_2 is a constant, x_i^T is the final opinion of user i, x_T is the target opinion on which we want the users to reach agreement. The terminal cost $\mathcal{G}_i$ denotes the distance between user i's target opinion and the desired opinion. Therefore, by defining the cost functional as (3.6), we want to achieve the target that all the users can reach an agreement on the desired opinion with as less efforts as possible. There are many real-life scenarios of reaching a consensus. For instance, during a presidential election, a candidate needs to implement an appropriate political policy (the control u_i) so that the voters can reach an agreement on voting for her (the target opinion x_T).

In summary, the optimal control problem faced by user i is given by

$$\inf_{u_i \in \mathcal{U}_i} \mathcal{J}_i(x_i, u_i), \tag{3.9}$$

$$\text{s.t. } dx_i(t) = f(x_i, x_i^{\mathcal{D}_i}, u_i)dt.$$

To this end, the belief and opinion evolution for a large number of SN users is formulated as a non-cooperative differential game.

3.1.3 High-Dimensional Mean Field Game

As the number of users N grows large, the complexity of solving the differential game in (3.9) can increase significantly due to the need to separately deal with the increasing interactions. Therefore, we propose the high-dimensional MFG reformulation of (3.9). Under the framework of MFG, we consider the collective behavior of all users instead of separately dealing with each user as shown in Fig. 3.1. In particular, the "mean field" in this work refers to the probability distribution of opinions. Now we can drop the index i because users are indistinguishable in MFG and we only focus on the collective behavior of them. Let $\rho(x, t)$ denote the probability of opinion x at time t, then the cost functional (3.6) is now transformed into

$$\begin{aligned} \mathcal{J}(\rho, u) &= \int_0^T \left\{ \int_\Omega c_1 ||u(t)||^2 \rho(x, t) dx \right\} dt + \int_\Omega c_2 ||x - x_0||_2 \rho(x, T) dx, \\ &= \int_0^T \left\{ \int_\Omega \rho(x, t) L(x, u) dx \right\} dt + \mathcal{G}(\rho(\cdot, T)), \end{aligned} \tag{3.10}$$

where Ω is the state space containing all possible opinions. For a generic user in a general asymmetric social network, the ratio between the average opinion of neighbors and the average opinion of all SN users is denoted as $r(x(t), t)$. This ratio is determined by the user's opinion because the social connections (neighborhood) of a person are very likely to be influenced by his/her opinion. For instance, when the opinion of a person is always extreme and unreasonable, then there will be fewer people who are willing to be his/her friends, i.e., the ratio r is very large. For a completely symmetric social network, where every SN user is connected (usually happens in a relatively small community), the ratio will be $r = 1$. In our model, the ratio function is a given function and needs to be estimated based on the specific SN. With this ratio function, the opinion dynamics in (3.5) are transformed into

$$\begin{aligned} dx(t) &= \left((1-\lambda)r(x,t)\int y\cdot\rho(y,t)dy + (\lambda-1)x(t) + \beta u(t)\right)dt + \sigma dw(t) \\ &= f(x(t),\rho(x,t),u(t))dt + \sigma dw(t), \end{aligned} \tag{3.11}$$

where λ is the stubbornness of a user in the population, $w(t)$ is the standard Wiener process that can capture opinion change due to random social events, σ is the diffusion constant, and y is a dummy variable to distinguish from x. For instance, people's opinions on the public health can be changed significantly due to the outbreak of COVID-19. In this case, the unexpected outbreak of COVID-19 is the random social event with a large diffusion constant σ. (3.11) can be represented in terms of the mean field $\rho(x,t)$ and then will be equivalent to the FP equation [186] given by

$$\partial_t\rho - \nu\Delta\rho + \nabla\cdot(\rho f) = 0, \tag{3.12}$$

where $\nu = \frac{\sigma^2}{2}$ is a constant. With cost functional (3.10) and the FP equation in (3.12), the high-dimensional MFG which describes the opinion evolution of a large number of SN users is summarized as

$$\begin{aligned} &\inf_{\rho,u} \mathcal{J}(x(t),\rho(x,t),u(t)), \\ &\text{s.t. } \partial_t\rho - \nu\Delta\rho + \nabla\cdot(\rho f) = 0, \\ &\rho(x,0) = \rho_0(x), \end{aligned} \tag{3.13}$$

where $u : \mathbb{R}^n \times [0,T] \to \mathbb{R}^n$ and ρ_0 is the initial probability distribution of SN users' opinions. To this end, the belief and opinion for a large number of SN users is formulated as a high-dimensional MFG.

Remark 3.1 To solve the problem in (3.9), user i needs to consider all neighbors' opinions $x^{\mathcal{D}_i}$. Separately dealing with every other neighbor's opinion makes the differential game in (3.9) too complex to solve. Therefore, we transform (3.9) into the MFG (3.13), where we only needs to consider the collective effect of $x^{\mathcal{D}_i}$. In particular, the influence of $x^{\mathcal{D}_i}$ is approximated by the probability distribution function $\rho(x, t)$ of users' opinions and the ratio function $r(x, t)$, given that the number of SN users is very large. The ratio function is a given function in our model and needs to be estimated based on the specific social network.

3.2 Generative Adversarial Networks Approach

In this section, we propose a GAN-based approach to solve the high-dimensional MFG in (3.13). In Sect. 3.2.1, we review the original GANs and show that they are minimax games with the goal of minimizing the distance between the real distribution and the generated distribution. In Sect. 3.2.2, we first establish conceptual connections between general MFGs and GANs. Then we show the mathematical connection to GANs for the high-dimensional MFG (3.13) by regarding it as a max-min game. Finally, in Sect. 3.2.3, we propose the GAN-like neural network for opinion evolution in SNs.

3.2.1 Generative Adversarial Networks

GANs, introduced by Goodfellow et al. [80] in 2004, are generative models that are designed to approximate an unknown and usually high-dimensional probability distribution P_r by generating appropriate parameterized probability distribution P_θ. Facing the difficulties of approximating many intractable probabilistic computations and leveraging the benefits of piecewise linear units, GANs used a framework consisting of two adversarial neural networks (NN): a generator network G_θ and a discriminator D_ω, where θ and ω are corresponding parameters. The generator is used to generate fake samples that have a similar distribution to P_r from a given latent space $\mathcal{Z}$, i.e., $G_\theta : \mathcal{Z} \to \mathcal{X}$. Meanwhile, the discriminator is trying to distinguish the generated fake samples from the real samples by showing the probability that the input sample is real, i.e., $D_\omega : \mathcal{X} \to [0, 1]$.Their training procedures take place simultaneously: the generator is trying to maximize the probability of D_ω making mistakes while the discriminator is trying to estimate the probability that a sample comes from the real distribution P_r instead of the generator. The training process ends when the discriminator can no longer distinguish the fakes samples from the real samples, i.e., the function D_ω equals $\frac{1}{2}$.

Mathematically, the goal of the original GAN is to find an equilibrium to the minimax game

$$\inf_{G_\theta} \sup_{D_\omega} \mathbb{E}_{x \sim P_r}[\log D_\omega(x)] + E_{z \sim \mathcal{Z}}[\log(1 - D_\omega(G_\theta(z)))]. \tag{3.14}$$

Keeping G_θ fixed and optimizing for D_ω in (3.14), the optimal discriminator is

$$D_G^*(x) = \frac{p_r(x)}{p_r(x) + p_\theta(x)}, \tag{3.15}$$

where p_r and p_θ are density functions of P_r and P_θ, respectively. Substituting the D_ω with (3.15), we get

$$\begin{aligned} &\inf_{G_\theta} \mathbb{E}_{x\sim p_r}\left[\log \frac{p_r(x)}{p_r(x) + p_\theta(x)}\right] + \mathbb{E}_{x\sim p_\theta}\left[\log \frac{p_\theta(x)}{p_r(x) + p_\theta(x)}\right] \\ &= -\log 4 + 2 \cdot JS(P_r, P_\theta). \end{aligned} \tag{3.16}$$

Regarding (3.16), the goal of training GANs is to minimize the Jensen-Shannon divergence between P_r and P_θ.

However, in practice, the updates of the generator in the original GAN with JS divergence get consistently worse as the discriminator gets better and the optimization becomes extremely unstable due to the discontinuous distributions of P_θ [6]. Therefore, [7] introduced Wasserstein GAN where the Wasserstein-1 distance is used to compute the distance between P_r and P_θ. The cost function (3.14) is then replaced by

$$\begin{aligned} \inf_{G_\theta} \sup_{D_\omega} \quad & \mathbb{E}_{x\sim P_r}[D_\omega(x)] - E_{z\sim Z}[D_\omega(G_\theta(z))], \\ \text{s.t.} \quad & \|\nabla D\| \leqslant 1, \end{aligned} \tag{3.17}$$

where the constraint requires that the discriminator to be 1-Lipschitz.

3.2.2 *Connections Between MFGs and GANs*

In this subsection, we first establish conceptual connections between MFGs and GANs. Then we show the underlying primal-dual structure, which is explicitly connected to Wasserstein GANs, of the high-dimensional MFG in (3.13).

3.2.2.1 MFGs as GANs

MFGs can be represented by the system of partial differential equations (PDEs) [117],

$$\begin{aligned} &\partial_t \phi - \nu \Delta \phi + H(x, \nabla \phi) = 0, \quad \text{(HJB)} \\ &\partial_t \rho - \nu \Delta \rho - \nabla \cdot (\rho \nabla_p H(x, \nabla \phi)) = 0, \quad \text{(FP)} \\ &\rho(x, 0) = \rho_0, \phi(x, T) = g(x, \rho(\cdot, T)), \end{aligned} \tag{3.18}$$

Table 3.1 Conceptual connections of MFGs and GANs

	GANs	MFGs
Generator G_θ	NN for approximating real distribution $G_\theta : \mathcal{Z} \to \mathcal{X}$	NN for solving the FP equation
Characterization of P_r	Sample data	FP equation for consistency
Discriminator D_ω	NN measuring distance between P_θ and P_r	NN for computing residual from the HJB equation

where HJB equation computes the optimal strategy of a generic user and the FP equation describes the evolution of the population, $\phi : \mathbb{R}^n \times [0, T] \to \mathbb{R}$ refers to the value function, $H : \mathbb{R}^n \times \mathbb{R}^n \to \mathbb{R}$ is the Hamiltonian, $\rho(\cdot, t) \in \mathcal{P}(\mathbb{R}^n)$ is the distribution of the users at time t, and $g : \mathbb{R}^n \times \mathcal{P}(\mathbb{R}^n) \to \mathbb{R}$ is the terminal cost. Intuitively, the following connections between MFGs and GANs can be observed [25]:

- The latent space $\mathcal{Z}$ can be regarded as $\mathbb{R}^d$ and the samples of the latent variables can be drawn from the initial population distribution ρ_0.
- The generator G_θ is used to approximate the FP equation, i.e., it takes the latent variables z and the time t as the input features in order to estimate the user distributions $\rho(x, t)$ from $t = 0$ to $t = T$.
- Equilibrium states of the MFGs are not explicitly available but do exist, just like the true distribution P_r of GANs. Therefore, the discriminator D_ω is used to measure the distance between the estimated population distribution and the real population distribution by checking the residual from the HJB equation.

In summary, comparisons of the roles of generator and discriminator in MFGs and GANs are given in Table 3.1

3.2.2.2 MFG as a Max-Min Game

For the high-dimensional MFG in (3.13), we can see a more transparent connection to GAN by showing that MFGs can be regarded as a max-min game. Denote ϕ as the Lagrange multiplier (the Lagrange multiplier ϕ will eventually satisfy the HJB equation and then become the value function), we can put the differential constraint in (3.12) (FP equation) into the cost function in (3.10) to obtain

$$
\begin{aligned}
\sup_{\phi} \inf_{\rho,u} \int_0^T \left\{ \int_\Omega \rho(x,t) L(x,u) dx \right\} dt + \mathcal{G}(\rho(\cdot, T)) \\
- \int_0^T \int_\Omega \phi(x,t)(\partial_t \rho - \nu \Delta \rho + \nabla \cdot (\rho f)) dx dt.
\end{aligned}
\tag{3.19}
$$

Here we define Ω as the spatial domain, e.g. $\Omega \subset \mathbb{R}^n$. To show the connection to GANs more explicitly, we give the proposition as follows:

Proposition 3.1 (MFG as the Max-min Game) *The MFG in* (3.19) *is equivalent to the max-min game, where one player controls the value function* ϕ *and the other player controls the probability density function* ρ

$$\begin{aligned}\sup_{\phi}\inf_{\rho}\int_0^T &\left\{\int_{\Omega}\rho(x,t)(\partial_t\phi+\nu\Delta\phi+H(x,\nabla\phi,\rho,r))dx\right\}dt\\ &+\int_{\Omega}\phi(x,0)\rho_0(x)dx-\int_{\Omega}\phi(x,T)\rho(x,T)dx\\ &+\mathcal{G}(\rho(\cdot,T)),\end{aligned}\tag{3.20}$$

where the Hamiltonian H is given by

$$\begin{aligned}H(x,\nabla\phi,\rho,r)&=\inf_u\{\nabla\phi\cdot f+L(x,u)\}\\ &=\inf_u\left\{\nabla\phi((1-\lambda)r(x,t)\int_{\Omega}x\rho(t,x)dx\right.\\ &\left.+(\lambda-1)x+\beta u)+u^2\right\}\\ &=-\frac{\beta^2\nabla\phi^2}{4}+\nabla\phi\left((1-\lambda)r(x,t)\int_{\Omega}\rho x dx+(\lambda-1)x\right).\end{aligned}\tag{3.21}$$

The formulation in (3.20) is the foundation of our GAN-like neural network. From (3.20), we can see that the high-dimensional MFG in (3.13) is a max-min game similar to GANs in (3.14) and (3.17).

Proof The proof is shown by (3.22), (3.23), (3.24).

First, we rewrite the Lagrange dual problem in (3.19) as follows

$$\begin{aligned}&\sup_{\phi}\inf_{\rho,u}\int_0^T\left\{\int_{\Omega}\rho(x,t)L(x,u)dx\right\}dt+\mathcal{G}(\rho(\cdot,T))\\ &-\int_0^T\int_{\Omega}\phi(x,t)(\partial_t\rho-\nu\Delta\rho+\nabla\cdot(\rho f))dxdt,\\ &=\sup_{\phi}\inf_{\rho,u}\int_0^T\left\{\int_{\Omega}\rho(x,t)L(x,u)dx\right\}dt\\ &+\mathcal{G}(\rho(\cdot,T))-\int_0^T\int_{\Omega}\phi(x,t)\partial_t\rho-\nu\phi(x,t)\Delta\rho\\ &+\phi(x,t)\nabla\cdot(\rho f)dxdt.\end{aligned}\tag{3.22}$$

We then integrate by parts with respect to t for $\int_0^T \int_\Omega \phi(x,t)\partial_t \rho \, dxdt$, integrate by parts with respect to x for $\int_0^T \int_\Omega \nu\phi(x,t)\Delta\rho \, dxdt$, and $\int_0^T \int_\Omega \phi(x,t)\nabla \cdot (\rho f) \; dxdt$, and we obtain

$$
\begin{aligned}
&\sup_{\phi} \inf_{\rho,u} \int_0^T \left\{ \int_\Omega \rho(x,t)L(x,u)dx \right\} dt + \mathcal{G}(\rho(\cdot,T)) \\
&\quad - \int_\Omega \rho\phi(x,t)\Big|_0^T dx + \int_0^T \int_\Omega \partial_t\phi \cdot \rho dxdt \\
&+ \int_0^T \nu(\phi \cdot \nabla\rho - \nabla\phi \cdot \rho)\Big|_\Omega dt + \int_0^T \int_\Omega \nu\rho \cdot \Delta\phi dxdt \\
&\quad - \int_0^T \rho f \cdot \phi(x,t)\Big|_\Omega dt + \int_0^T \int_\Omega \rho f \cdot \nabla\phi dxdt.
\end{aligned}
\tag{3.23}
$$

We assume the zero flux condition for the continuity equation. Hence, we have

$$
\nu(\phi \cdot \nabla\rho - \nabla\phi \cdot \rho)\Big|_\Omega = 0, \quad \rho f \cdot \phi(x,t)\Big|_\Omega = 0.
$$

Plugging the results in (3.23), we obtain

$$
\begin{aligned}
&\sup_{\phi} \inf_{\rho,u} \int_0^T \left\{ \int_\Omega \rho(\partial_t\phi + \nu\Delta\phi + f \cdot \nabla\phi + L)dx \right\} dt + \mathcal{G}(\rho(\cdot,T)) \\
&+ \int_\Omega \phi(x,0)\rho_0(x)dx - \int_\Omega \phi(x,T)\rho(x,T)dx \\
&= \sup_{\phi} \inf_{\rho} \int_0^T \left\{ \int_\Omega \rho(\partial_t\phi + \nu\Delta\phi + \inf_u\{f \cdot \nabla\phi + L\})dx \right\} dt + \mathcal{G}(\rho(\cdot,T)) \\
&+ \int_\Omega \phi(x,0)\rho_0(x)dx - \int_\Omega \phi(x,T)\rho(x,T)dx \\
&= \sup_{\phi} \inf_{\rho} \int_0^T \left\{ \int_\Omega \rho(\partial_t\phi + \nu\Delta\phi + H(x,\nabla\phi,\rho,r))dx \right\} dt + \mathcal{G}(\rho(\cdot,T)) \\
&+ \int_\Omega \phi(x,0)\rho_0(x)dx - \int_\Omega \phi(x,T)\rho(x,T)dx, \\
&\text{where the Hamiltonian } H(x,\nabla\phi,\rho,r) = \inf_u\{f \cdot \nabla\phi + L\}.
\end{aligned}
\tag{3.24}
$$

In particular, a Wasserstein GAN can be regarded as a specific instance of a deterministic MFG [132]. Let $\sigma = 0$ (eliminate the effect of random phenomenon), let ρ_T be the target measure that can optimize the terminal cost $\mathcal{G}$. Then the FP equation resembles the optimal transport maps from a distribution ρ_0 to $\rho(\cdot, T)$ in Wasserstein GAN. Let H be the Hamiltonian defined by

$$H(x, p) = 1_{||p|| \leqslant 1} = \begin{cases} 0 & ||p|| \leqslant 1, \\ \infty & \text{otherwise}, \end{cases} \tag{3.25}$$

and then (3.20) is reduced to

$$\begin{aligned} \sup_{\phi} \int_{\Omega} \phi(x)\rho_0(x)dx - \int_{\Omega} \phi(x)\rho_T(x)dx, \\ \text{s.t.} \quad ||\nabla \phi(x)|| \leqslant 1. \end{aligned} \tag{3.26}$$

And since $H(p) = 1_{||p|| \leqslant 1}$, we have that $\partial_t \phi = 0$, and $\phi(x, t) = \phi(x)$ for all t. As $\partial_t \phi = 0$ and $||\nabla \phi(x)|| \leqslant 1$ hold, we have $\partial_t \phi + H(\nabla \phi) = 0$. We then can see that (3.26) is precisely the Wasserstein-1 distance.

3.2.3 GAN-Like Neural Network for Opinion Evolution

We solve (3.20) by training a GAN-like neural network. Detailed procedures of our approach are shown in Algorithm 1. First, we initialize the neural networks $N_\theta(z, t)$ (discriminator) and $G_\theta(z, t)$ (generator). Then we set

$$\begin{aligned} \phi_\omega(x, t) &= (1 - t)N_\omega(x, t) + tg(x), \\ G_\theta(z, t) &= (1 - t)z + tN_\theta(z, t), \end{aligned} \tag{3.27}$$

where $z \sim \rho_0$ are samples drawn from the initial distribution. The formulation of G_θ in (3.27) automatically encodes the initial distribution ρ_0 of the opinion. Moreover, this neural network encodes the underlying structure of MFGs via (3.20) and (3.27), which absolves the neural networks form learning the entire solution of the MFG from scratch.

During the training process, we alternatively training G_θ (the population distribution on the opinions), and ϕ_ω (the value function for each agent). Specifically, in order to train ϕ_ω, we first sample a batch $\{z_b\}_{b=1}^B$ from the given initial opinion distribution ρ_0, and $\{t_b\}_{b=1}^B$ uniformly from $[0, T]$. Then, we compute the out of the generator $x_b = G_\theta(z_b, t_b)$ for $b = 1, \cdots, B$. The total loss of the discriminator ϕ_ω is given by

$$
\begin{aligned}
\text{loss}_{\phi} = & \frac{1}{B}\sum_{b=1}^{B}\phi_{\omega}(x_b, 0) + \frac{1}{B}\sum_{b=1}^{B}\partial_t\phi_{\omega}(x_b, t_b) \\
& + \nu\Delta\phi_{\omega}(x_b, t_b) + H(\nabla_x\phi_{\omega}(x_b, t_b), x_b, r(x_b, t_b)),
\end{aligned} \tag{3.28}
$$

where the Hamiltonian $H(\nabla_x\phi_{\omega}(x_b, t_b), x_b)$ is given by

$$
\begin{aligned}
H(\nabla_x\phi_{\omega}(x_b, t_b), r(x_b, t_b), x_b) = & -\frac{\beta^2}{4}\nabla_x\phi_{\omega}(x_b, t_b)^2 \\
+ \nabla_x\phi_{\omega}(x_b, t_b)\Big((1-\lambda)r(x_b, t_b)\bar{x_b} + (\lambda - 1)x_b\Big).
\end{aligned} \tag{3.29}
$$

where $\bar{x_b}$ is the mean value of x_b and λ is the stubbornness.

We can optionally add a regularization term

$$
l_{\text{HJB}} = \gamma\frac{1}{B}\sum_{b=1}^{B}||\partial_t\phi_{\omega}(x_b, t_b) + \nu\Delta\phi_{\omega}(x_b, t_b) + H(\nabla_x\phi_{\omega}(x_b, t_b), r(x_b, t_b), x_b)|| \tag{3.30}
$$

to correct deviations from the HJB equations. This way of regularization also plays a role in the Wasserstein GAN, where the norm of the gradient is penalized. At last, we backpropagate the total loss to update the weights of the discriminator ϕ_{ω}. In order to train the generator, we also sample $\{z_b\}_{b=1}^{B}$ and $\{t_b\}_{b=1}^{B}$, and the loss of the generator is given by

$$
\begin{aligned}
\text{Loss}_G = & \frac{1}{B}\sum_{b=1}^{B}\partial_t\phi_{\omega}(G_{\theta}(z_b, t_b), t_b) + \nu\Delta\phi_{\omega}(G_{\theta}(z_b, t_b), t_b) \\
& + H(\nabla_x\phi_{\omega}(G_{\theta}(z_b, t_b), t_b), r(x_b, t_b), x_b).
\end{aligned} \tag{3.31}
$$

At last, we backpropagate the loss to update the weights of G_{θ}. In conclusion, in each time slot $t \in [0, T]$, the GAN-like neural network will be trained. The generator G_{θ} will generate the opinion distribution at time t and the discriminator will get the result of the value function at time t.

3.3 Simulation Results

3.3.1 Simulation Parameters

For the social network, we set $\beta = 1$ (the weight of the control in (3.5)), $c_1 = 0.5$ (the weight of the control effort in (3.7)), and $c_3 = 1$ (the weight of the terminal cost in (3.7)). The opinions of SN users are initialized as a Gaussian distribution

Algorithm 1 GAN-based approach for opinion evolution in social nets

Require: ν diffusion constant, H Hamiltonian
Require: Initialize neural networks N_ω and N_θ, batch size B
Require: Set ϕ_ω and G_θ as in (3.27)
while not converge **do**
  Train ϕ_ω:
  Sample batch $\{(z_b, t_b)\}_{b=1}^B$ where $z_b \sim \rho_0$ and $t_b \sim \text{Unif}(0, T)$
  $x_b \leftarrow G_\theta(z_b, t_b)$ for $b = 1, \cdots, B$
  $l_0 \leftarrow \frac{1}{B}\sum_{b=1}^B \phi_\omega(x_b, 0)$
  $l_t \leftarrow \frac{1}{B}\sum_{b=1}^B \partial_t\phi_\omega(x_b, t_b) + \nu\Delta\phi_\omega(x_b, t_b) + H(\nabla_x\phi_\omega(x_b, t_b), x_b, r(x_b, t_b), z_b)$
  $l_{\text{HJB}} \leftarrow \gamma\frac{1}{B}\sum_{b=1}^B ||\partial_t\phi_\omega(x_b, t_b) + \nu\Delta\phi_\omega(x_b, t_b) + H(\nabla_x\phi_\omega(x_b, t_b), x_b, r(x_b, t_b), z_b)||$
  Backpropagate the loss $l_{\text{total}} = l_0 + l_t + l_{\text{HJB}}$ to ω weights

  train G_θ:
  Sample batch $\{(z_b, t_b)\}_{b=1}^B$ where $z_b \sim \rho_0$ and $t_b \sim \text{Unif}(0, T)$
  $l_t \leftarrow \frac{1}{B}\sum_{b=1}^B \partial_t\phi_\omega(G_\theta(z_b, t_b), t_b) + \nu\Delta\phi_\omega(G_\theta(z_b, t_b), t_b)$
 $+H(\nabla_x\phi_\omega(G_\theta(z_b, t_b), t_b), x_b, r(x_b, t_b), z_b)$
  Backpropagate the loss $l_{\text{total}} = l_t$ to θ weights
end while

centering at $(-1, 0)$ and the target opinion X_T is set at $(1, 0)$. Residual neural networks (ResNets) are used for both networks. $Tanh$ activation function is used in ϕ_ω while $ReLU$ activation function is used in G_θ. We choose ADAM as the optimizer with $\hat{\beta} = (0.5, 0.9)$, the learning rate 4×10^{-4} for ϕ_ω when $\nu = 0$, the learning rate 10^{-4} for ϕ_ω when $\nu = 0.4$, the learning rate 1×10^{-4} for G_θ when $\nu = 0$ and $\nu = 0.4$.

3.3.2 *Effect of Stochasticity and Convergence Analysis*

Guided by the Hegselmann-Krause opinion dynamics in (3.1), where a generic user is trying to revise his/her opinions based on the average opinion of his/her neighbors, users can reach an agreement very quickly even before they get the target opinion, as shown in Fig. 3.2a–c. Furthermore, the lower the dimensions of users' opinions are, the faster they can reach an agreement. In other words, reaching an agreement on fewer social events is easier for a group of SN users.

As shown in Fig. 3.2e–f, by setting the diffusion constant $\nu = 0.4$, we can observe the effect of random events on users' opinions. Compared with Fig. 3.2a–c, we figure out that users can not form a dense distribution in the opinion space, i.e., reach an agreement easily. However, the cone shape of users' distributions can imply the trend of agreements. Moreover, when reaching the target opinion, an agreement can be achieved regardless of the effect of random events.

In Fig. 3.3a, b, the convergence of the GAN-like neural network is shown by the convergence of logarithm of the HJB residual error defined in (3.30). The neural network can converge around 100,000 epochs and can solve the MFG up

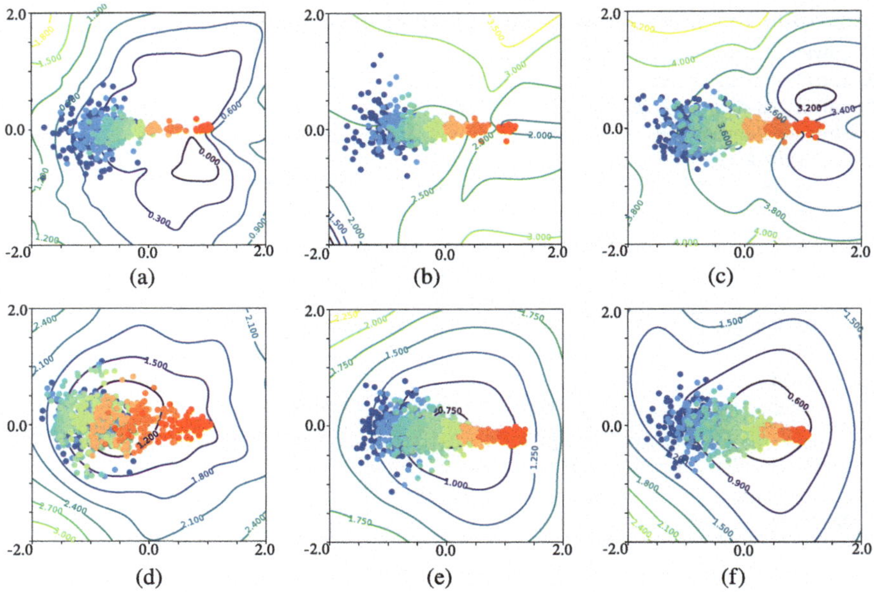

Fig. 3.2 Plots of SN users' distributions in the opinion space. (**a**) $\nu = 0$,dim=2. (**b**) $\nu = 0$,dim=50. (**c**) $\nu = 0$,dim=100. (**d**) $\nu = 0.4$,dim=2. (**e**) $\nu = 0.4$,dim=50. (**f**) $\nu = 0.4$,dim=100

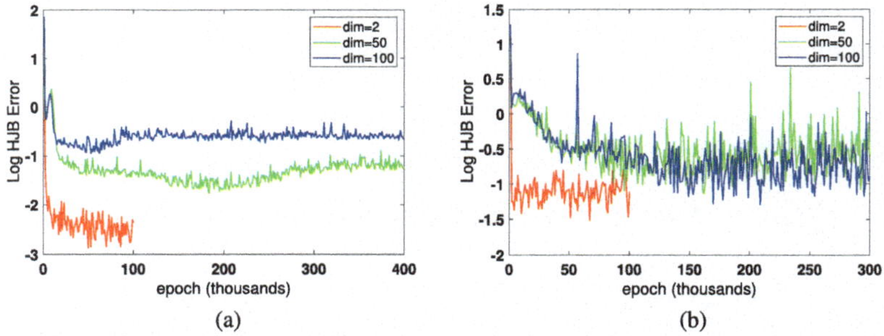

Fig. 3.3 Convergence analysis. (**a**) $\nu = 0$. (**b**) $\nu = 0.4$

to 100 dimensions. We have a larger error in higher dimensions. The solutions after convergence, like other machine-learning-based techniques, depend on the setting of the parameters of the neural networks.

3.3.3 Effect of Users' Stubbornness

Integrating the concept of stubbornness in the FJ model to the HK model, users' opinions are revised according to the dynamics in (3.3), where new opinions of a

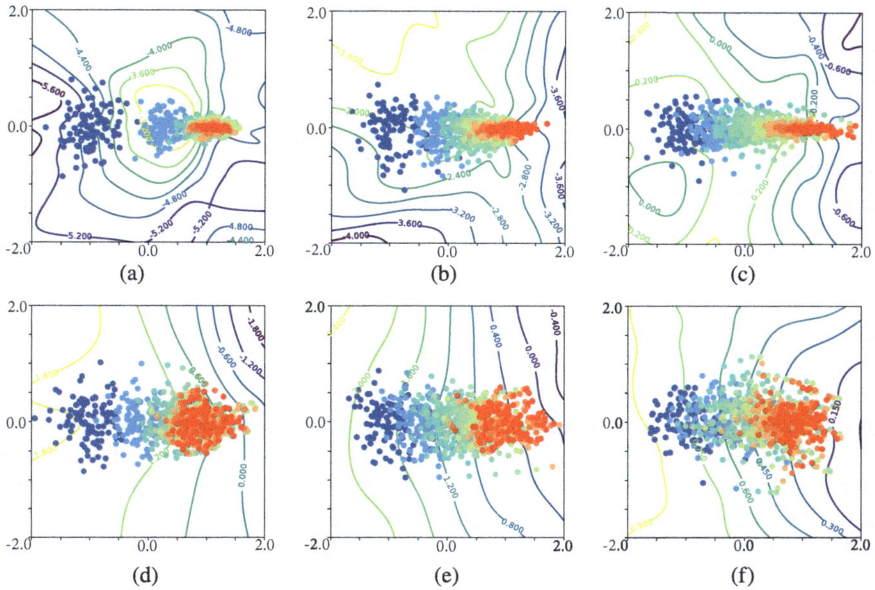

Fig. 3.4 Effect of users' stubbornness (λ) on the opinion evolution. (**a**) $\nu = 0, \lambda = 0.2$. (**b**) $\nu = 0, \lambda = 0.5$ (**c**) $\nu = 0, \lambda = 0.8$. (**d**) $\nu = 0.4, \lambda = 0.2$. (**e**) $\nu = 0.4, \lambda = 0.5$. (**f**) $\nu = 0.4, \lambda = 0.8$

generic user i are determined by the average opinions of her neighbors, her current opinions and her control efforts. The weight of the current opinion λ_i represents the strength of stubbornness of user i. In Fig. 3.4, opinions of 100 SN users between $t = 0$ and $t = 1$ are depicted. We conduct the experiments with three different strengths of stubbornness ($\lambda = 0.2$, $\lambda = 0.5$, and$\lambda = 0.8$) under the influence of random social issues.

Comparing Fig. 3.4a with Fig. 3.2b, where there is no stubbornness, it is more difficult to form a congested opinion distribution with stubbornness $\lambda = 0.2$, i.e., stubbornness can reduce the speed of reaching an agreement. If SN users do not insist on their initial opinions ($\lambda = 0$) and actively revise their opinions according to their neighbors' opinions, the first agreement appears around $t = 0.625$ as shown in Fig. 3.2b. With $\lambda = 0.2$, the first agreement appears around $t = 1$ and final opinions are more dispersed compared with non-stubborn case.

Comparing Fig. 3.4a, c, we can also figure out that the speed of reaching an agreement is reduced by stubbornness. In Fig. 3.4a, opinions at $t = 0.125$ center around $(0, 0)$ and show an obvious phenomenon of congestion compared with the initial opinions. In Fig. 3.4c, opinions just slightly deviate from the initial opinions. As the strength of stubbornness continues to increase, we find that users cannot reach an agreement at $t = 1$. The distribution of the red points becomes more and more scattering between the initial opinion and the target opinions as the λ increases from 0.2 to 0.8. In other words, an agreement cannot form when the strength of stubbornness reaches a certain level.

Finally, we analyze the effect of users' stubbornness under the influence of random social issues ($\nu = 0.4$). Comparing Fig. 3.4a, d, we can see that opinions become rather dispersed under the influence of stochasticity. In Fig. 3.4a, SN users can reach an agreement at $t = 1$ though the speed is reduced. In Fig. 3.4d, when there exists the effect of random social issues, stubbornness becomes much more powerful in stopping the users from reaching an agreement. The same phenomenon can be observed by comparing Fig. 3.4b with Fig. 3.4e or comparing Fig. 3.4c with Fig. 3.4f.

3.3.4 Modeling Opinion Evolution with Real Data Set

In this subsection, we model the opinion evolution with the Social Evolution data set [142]. In Sect. 3.3.4.1, we introduce the data set released by the MIT Human Dynamics Lab. In Sect. 3.3.4.2, we define the evaluation metrics used to show the goodness of estimation. In Sect. 3.3.4.3, we show the procedures of our experiment and analyze the performance of the GAN-like neural network.

3.3.4.1 Data Set

We use the Social Evolution data set, which contains surveys and sensor data about the diffusion of music preferences, political opinions, diet, exercise, eating habits, and epidemiological contagion through the interpersonal relationships of about 70 residents (around 80% of the total population) of an undergraduate dormitory in MIT. In our experiment, we use the statistics regarding the social evolution of music preferences between September 2008 and November 2008. Students' opinions on heavy metal music were surveyed and divided into four categories: no interest, slight interest, medium interest, and high interest.

3.3.4.2 Evaluation Metrics

The coefficient of determination (R^2) and average relative bias are used to evaluate the goodness of estimation. R^2 can indicate the proportionate amount of variation in the real data value y_i explained by the estimated value $\hat{y}_i$. The ordinary R^2 is computed by

$$R^2 = 1 - \frac{\sum_{i=1}^{n} \omega_i (y_i - \hat{y}_i)^2}{\sum_{i=1}^{n} \omega_i (y_i - \bar{y})^2}, \tag{3.32}$$

where $\bar{y}$ is the mean of real data value y_i. The average relative bias is computed by

$$Bias = \frac{1}{n}\sum_{i=1}^{n}\frac{|y_i - \hat{y}_i|}{|y_i|}. \tag{3.33}$$

3.3.4.3 Procedures of Experiment and Performance Analysis

First, we need to process the statistics regarding the participants' opinions on heavy metal music to use the GAN-like neural network to model the opinion evolution. 74 residents provided information about their interests in heavy metal music during September 2008 while 71 residents updated this information during November 2008. We counted the number of residents in the four categories and obtained the discrete probability distribution of the population. Then we approximate the discrete population distribution with a continuous Gaussian distribution. To obtain the best approximation, we assign different values to different opinions as follows: "medium interest"= -1.5, "slight interest"= -1.0, "no interest"= 0.2, and "high interest"= 1.6.

The approximated Gaussian distribution of data collected in September 2008 serves as ρ_0 and is used to generate the latent variable z_b. Meanwhile, the time t_b is sampled from a uniform distribution between $t = 0$ and $t = 1$. Combining z_b and t_b, we get the complete latent variable for the generator. The stubbornness λ is set to 0.2 while the stochasticity parameter ν is set to 0.4. The target opinion is set to $(-1.5, -1.5)$. It should be mentioned that it will be impossible for the participants to reach an agreement on medium interest in heavy metal music, either due to students' stubbornness or the effect of random social issues. However, a target opinion can set up a general trend for the opinion evolution. After setting related parameters, the neural network is trained to estimate the opinions distribution in November 2008.

In Fig. 3.5, we show the prediction of our GAN-like neural network. The blue bar represents the raw data collected from the participants in November 2008, the red curve is the Gaussian distribution approximating the data, and the black dotted curve is the estimation given by the neural network. The average relative bias is 6.81% meaning that the estimation slightly deviates from the data. The coefficient of determination is 0.9745 meaning that the variance of the data is well represented by the estimation. Therefore, we conclude that the proposed neural network achieves a good prediction on the opinion evolution of students' music preferences.

3.4 Chapter Summary

Studying the diffusion of opinions and ideas in SNs can not only help us understand how people make decisions and reach a consensus but also can provide a solid foundation for many valuable social applications. To avoid separately dealing with interactions between a large number of SN users, we have formulated the problem of opinion evolution as a high-dimensional MFG. Suffering from the curse of dimensionality, existing numerical methods are not able to solve the proposed high-dimensional MFG. Therefore, we have applied the GAN-like neural network to efficiently solve the stochastic MFG of up to 100 dimensions. In simulations, we have analyzed the effects of random social issues and stubbornness on opinion evolution. Finally, with a real data set as the benchmark, it has been shown that the proposed algorithm can efficiently estimate and predict the diffusion of opinions in SNs.

Appendix: Effect of Opinion Dynamics on MFG

In this appendix, we show how opinion dynamics affect the MFG. In particular, we show the non-convexity of the MFG formulation (3.13) caused by the opinion diffusion term $f(x(t), \rho(x,t), u(t))$ in (3.11). To prove the non-convexity of (3.13), we need to show the running cost ((3.7)) is non-convex when the differential constraint (3.12) is linear to both ρ and u.

Observing the FP equation (3.12), we can find that it is not linear to both ρ and u due to the product of ρf in the divergence operator. Therefore, we need to define a flux $m = \rho f$, which is the product of density function ρ and diffusion vector field f. Moreover, the above change of variable is a common practice when the FP equation is the differential constraint [13]. With this change of variable, we make the FP equation linear to both the flux m and the density ρ, which can be given by

$$\partial_t \rho - \nu \Delta \rho + \nabla \cdot m = 0. \tag{3.34}$$

Correspondingly, we need to show that the running cost is not convex in ρ and m. Therefore, we first represent u with m, which is given by

$$u = \frac{m}{\rho} + (\boldsymbol{\lambda} - \mathbf{1})\boldsymbol{r} \int \boldsymbol{y} \cdot \boldsymbol{\rho}(\boldsymbol{y}, \boldsymbol{t}) \boldsymbol{dy} + (\mathbf{1} - \boldsymbol{\lambda})\boldsymbol{x}, \tag{3.35}$$

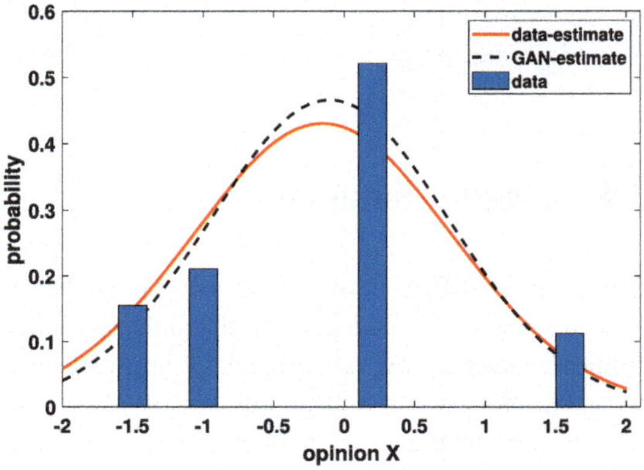

Fig. 3.5 Estimation of opinions distribution in November, 2008

where we set the constant $\beta = 1$ without loss of generality. The diffusion caused by the opinion dynamics is highlighted in bold text. Then we substitute u with (3.35) in (3.7) and we obtain

$$
\begin{aligned}
L(m, \rho) &= \frac{1}{2}\rho||u||^2 \\
&= \frac{1}{2}\rho||\frac{m}{\rho} + (\lambda - 1)r \int y \cdot \rho(y, t)dy + (1 - \lambda)x||^2 \\
&= \boldsymbol{m \cdot \left((\lambda - 1)r \int y \cdot \rho(y, t)dy + (1 - \lambda)x\right)} \\
&+ \frac{1}{2}\frac{||m||^2}{\rho} + \frac{\rho}{2}||(\lambda - 1)r \int y \cdot \rho(\cdot, t)dy + (1 - \lambda)x||^2,
\end{aligned}
$$

where we find that the highlighted part is not convex in ρ and m due to the product of the integral of ρ and m.

In summary, the inclusion of the opinion dynamics in the state transition function makes the optimal control problem in (3.13) non-convex.

Chapter 4
Data Augmentation Using Mean Field Games

Data augmentation is a critical component in building modern deep-learning systems. In this chapter, we propose `MFG Augment`, a novel data augmentation method based on the Mean-Field-Game (MFG) theory, that can synthesize a sequence of data between every two images or features. The central idea is to consider every image as a distribution over its pixel or feature space. Using Mean-field Game theory, we can generate a time-continuous "path" from one distribution to another so that the points along the "path" are augmented images or features. Empirically, the experiment results on MNIST, CIFAR-10, and ImageNet demonstrate that the proposed technology has better generalization ability and higher classification accuracy as compared to several benchmark methods. More importantly, our `MFG Augment` improves the test accuracy significantly when the dataset size is small. MFG Augment consistently shows better affinity and diversity scores, two important empirical metrics for evaluating the generalization of data augmentation techniques.

4.1 Introduction

Data augmentation is proven to be a crucial technique for performing various deep learning tasks [182, 194], such as image classification [44, 45, 55, 78], speech recognition [155], natural language processing [53], etc. It aims to increase the amount of training data by adding slightly modified copies of existing data or newly created synthetic data from existing data. This mechanism helps learning models be more robust to common data variation. In the image domain, popular augmentation methods include translating the image by part of its pixels or flipping and rotating the whole image [41, 55, 88, 166, 204]. In the feature domain, common augmentations include interpolating learned features with linear transformation, adding noise, interpolating, and extrapolating between them [54, 188], or utilizing a deep neural network to model the feature trajectories [135, 152].

Y. Kang et al., *Mean Field Guided Machine Learning*, Wireless Networks,
https://doi.org/10.1007/978-3-031-91859-9_4

On the other hand, the Mean-field Game theory, an advanced variant of the optimal transport theory, has been proposed by Lasry and Lions in [117] and Caines and Huang in [22] in recent years. It is a mathematical framework developed to analyze situations where a large number of participants (or agents) make decisions based on their individual circumstances while considering the average behavior of the entire group. Here, the agents in the proposed `MFG Augment` refer to the pixels or the learned feature of an image. The basic idea of the mean-field game is that it can control the movement of the distribution of a large number of agents and transform it into a target distribution along an optimized "path". Different from traditional data augmentation methods that typically employ heuristic techniques, such as rotation, flipping, or cropping, without a unifying theoretical framework to guide and validate the quality and relevance of the generated data [198], `MFG Augment` offers a systematic and theoretical foundation to generate new data by modeling the relationship between original and augmented data distribution. Since its inception, MFG has achieved many successful applications in computer vision, manifold learning, reinforcement learning innovations, etc. For example, the famous Wasserstein Generative adversarial neural network (W-GAN) is based on the MFG theory to overcome the vanishing gradient problem [8]. It has also been used to find discriminant or robust subspaces for a dataset [65, 156]. Moreover, MFG theory can guide color transfer between images [64] or find correspondences between languages in word embeddings [3]. Other applications of MFG theory in engineering include robot swarm control [103], age of information minimization in wireless communication networks [212], resource allocation [125], data offloading optimization in edge computing [105], etc.

In this chapter, we propose `MFG Augment`, a novel data augmentation method based on the MFG theory. We formulate the data augmentation process into a quantifiable MFG optimization problem in either the image's pixel space or feature space, and our approach can generate both augmented images and features (Fig. 4.1). Examples of image-level `MFG Augment` and feature-level `MFG Augment` are shown in Figs. 4.2 and 4.3, respectively. To the best of our knowledge, this is the first work that integrates the Mean-field Game theory to data augmentation.

We demonstrate that the proposed `MFG Augment` has three main advantages: (1) The image-level MFG Augment is able to keep the shape and edge of the objects in images. (2) The feature-level augmentation is able to generate features in the learned manifold under quantifiable optimality conditions. (3) Empirically, our `MFG Augment` can improve the test accuracy significantly, especially for small datasets, improving from AugMix's 48.55 training samples from CIFAR-10. Contributions of this work are summarized as follows.

- We formulate the data augmentation process as an MFG model based on the MFG theory framework, where the agent is modeled as either the pixel or the learned feature of an image. Based on the MFG model, we formulate the data augmentation process into an MFG optimization problem.

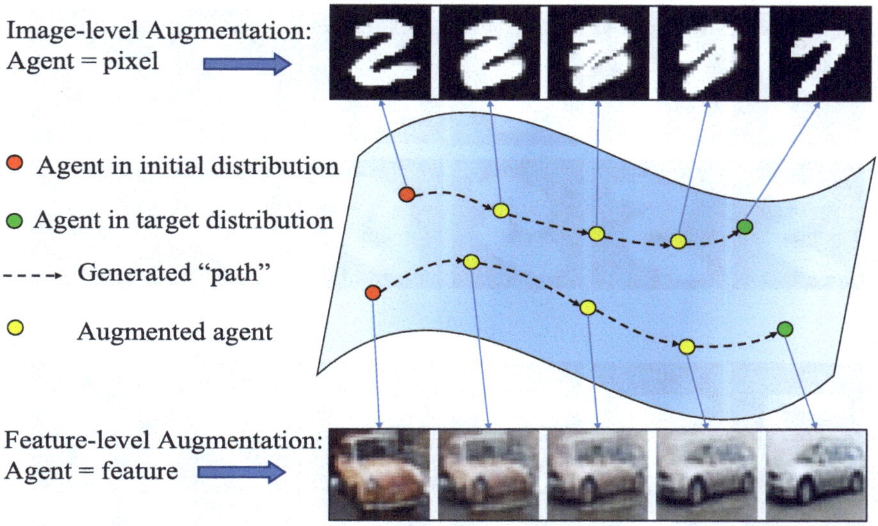

Fig. 4.1 Illustration of our `MFG Augment`: idea 1: image-level augmentation: every image is regarded as a distribution over its pixel value space. The upper red point represents an image's pixel distribution. We use `MFG Augment` to transform from one image's distribution (red point) to another image's pixel distribution along a "path" (dotted line) so that the points along the path (yellow points) are augmented images. Idea 2: feature-level augmentation: the whole dataset is regarded as a distribution over its feature space. We use `MFG Augment` to transform a distribution of a set of features (red point) to another distribution of a set of features (green point) along a path "dotted line", so that the points along the path (yellow points) are set of augmented features

- To obtain the optimal augmentation strategy, we propose a G-prox PDHG algorithm and a GAN-like network to solve the proposed MFG data augmentation problem.
- We present a comprehensive analysis of the proposed `MFG Augment`, including optimality and time complexity analysis, all underpinned by the rigorous and theoretically-informed MFG theory.
- Experiments implemented on existing benchmark datasets demonstrate a significant improvement in test accuracy over state-of-the-art methods.

The empirical results on image classification tasks show that our method achieves significant improvements over existing data augmentation techniques, including `Cutout` [55], `CutMix` [204] and `Augmix` [88], on databases: MNIST, CIFAR-10, and ImageNet. In Sect. 4.2, we provide an extensive literature review in the field of data augmentation. Then we propose `MFG Augment` in Sect. 4.3. In Sect. 4.4, we give a comprehensive analysis of the proposed `MFG Augment`. In Sect. 4.5, we show the experiment results of the proposed `MFG Augment` by comparing them with several benchmarks. Finally, conclusions and future works are summarized in Sect. 4.6.

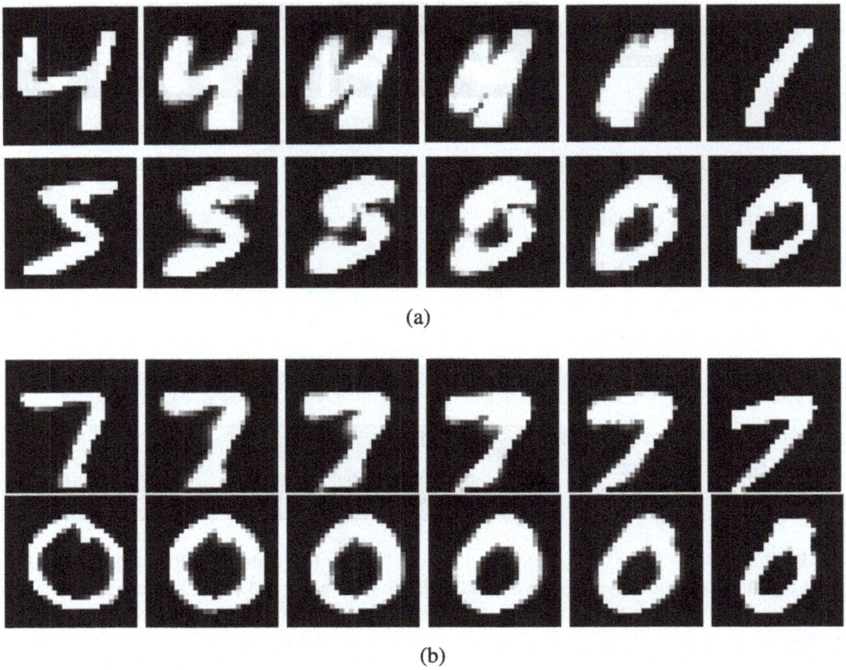

Fig. 4.2 Examples MFG augment in image-level augmentation on MNIST. (**a**) Examples of label-variant `MFG Augment` in image-level augmentation. (Transform a “4” into “1”, a “5” into “0”). (**b**) Examples of label-agnostic `MFG Augment` in image-level augmentation. (Transform a “7” into another “7”, a “0” into another “0”)

4.2 Related Work

Data augmentation stands as a cornerstone technique in enhancing the performance and generalization of machine learning models. Historically, data augmentation methods can be broadly categorized into two main classes based on the level at which they operate: image-level and feature-level. This categorization not only helps in understanding the evolution and diversity of augmentation techniques but also contextualizes their application based on specific needs and challenges. We will discuss these two categories in Sects. 4.2.1 and 4.2.2, respectively.

4.2.1 Image-Level Augmentations

Rotation, scale, translation, flip, elastic distortion, and jittering are common techniques used to augment image data [41, 166, 173, 198]. Random blocking techniques such as `Cutout` [55] that randomly masks out square regions of input images during training, `Hide-and-Seek` [174] that hides patches in a training image randomly,

(a)

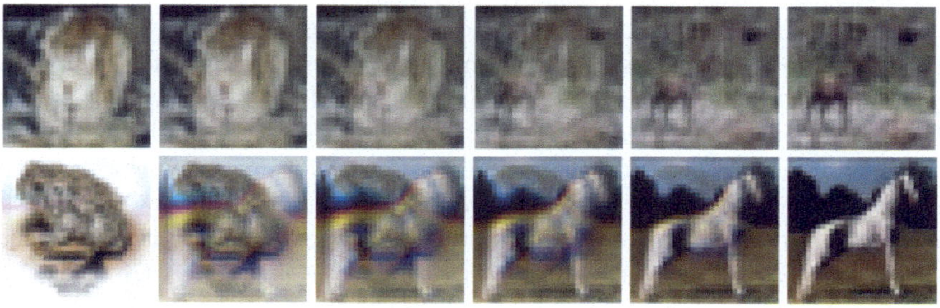

(b)

Fig. 4.3 Visualization examples of MFG Augment in feature-level augmentation on CIFAR-10 (Decoded augmented features). (**a**) Examples of label-agnostic `MFG Augment` in feature-level augmentation. (Transform a bird into another bird, a car into another car). (**b**) Examples of label-variant `MFG Augment` in feature-level augmentation (Transform a cat into a deer, a frog into a horse)

`Grid Mask` [31] that applies a grid-patterned mask comprised of regularly spaced rectangular regions onto input images, and `Random Erasing` [217] that involves erasing, or replacing, a randomly selected rectangular region within an input image, can also improve accuracy on clean data. Instead of randomly erasing images pixels, the authors in [78] use the saliency map to detect important regions on the original images and then preserve these informative regions during augmentation, which allows us to generate more faithful training examples. Rather than augmenting one image, some other techniques merge multiple input instances. For example, `Cutmix` [204] operates by cutting and pasting patches between two images and adjusting their corresponding labels proportionally. `Mixup` [210] combines two images using a convex combination. `SampleParing` [96] synthesizes a new sample from one image by overlaying another image. `BC Learning` [183] generates between-class images by mixing two images belonging to different classes with a random ratio. Also, `Co-Mixup` [109] maximizes the salient signal of input images and diversity among the augmented images. `CutPas` [59], together with some of its variants such as `Simple CutPas` [74] and `Continuous CutPas` [197], cuts object instances and paste them onto random backgrounds. Similarly, `FMix` [84] is also a mixed sample data augmentation technique that uses random binary masks obtained by applying a threshold to low-frequency images sampled from Fourier space. SuperMix [48] optimizes a mask for fusing two images to exploit the salient region with the Newton iterative method, which is 65x faster than gradient descent. Scale and Blend [73] cuts and scales object instances, and blends them in meaningful locations. Context DA [58] merges object instances using context guidance. ClassMis is proposed in [153] that generates augmentations by mixing unlabelled samples, by leveraging on the network's predictions for respecting object boundaries. Differently, the authors in [175] claim that non-label-preserving data augmentation can be surprisingly effective in improving model performance, and thus propose to use a non-linear combination of images to create augmented images.

Until now, the discussed augmentation techniques are model-free methods. However, various model-based data augmentation approaches, which utilize pre-trained models for generating augmented images, have likewise shown considerable effectiveness. For instance, in [5, 67, 180], a generative model is trained to do data augmentation, and in [143] the authors propose BAGAN as an augmentation tool to restore balance in imbalanced datasets. In [162], the authors train a CycleGAN to transform contrast CT images into non-contrast images, and then use the trained CycleGAN to augment training using these synthetic non-contrast images. BDA [184] uses CGAN to generate images using the Monte Carlo algorithm. MFC-GAN [2] handles the class imbalance problem by using multiple fake classes to obtain a fine-grained image for the minority classes. ImbCGAN [57] approximates the true data distribution and generate data for the minority classes of various imbalanced datasets. AugGAN [94] semantically preserves object when changing its style. Other types of data augmentation methods include using images with transferred style as in [147] that allows to generate the new images of high perceptual quality that combine the content of a base image with the appearance of another ones, and StyleMix [97] merges two images with style, content, and labels.

Besides those pixel modification methods, some methods learned augmentation policies, such as AutoAugment [44]. These methods search a group of augmentation operations to determine the optimal ones that optimize the downstream task's performance. A variant of AutoAugment named Fast AutoAugment is proposed to speed up the searching process using efficient density matching for augmentation policy search, Faster AutoAugment uses a differentiable policy search pipeline via approximate gradients, and Adversarial AutoAugment [211] that simultaneously optimizes target related object and augmentation policy search loss. Also, Randaugment is proposed in [45] that reduces the searching space that jointly controls all operations and hence reduces the computational expense of automated augmentation. PBA [89] adopts non-stationary augmentation policy schedules instead of a fixed augmentation policy via population-based training. LSSP [129] presents a policy-driven sequential image augmentation approach for image-related tasks. AdaTransform [181] uses a competitive task to obtain augmented images, in the training stage, a competitive task is employed to acquire augmented images with high task loss, and in the testing stage, a cooperative task is utilized to generate augmented images with low task loss. CDST-DA [159] utilizes a GAN to optimize a generative sequence, ensuring the transformed image aligns with the same class distribution. In [119], the authors use an influence function to predict how validation loss is affected by a particular training sample and minimize the approximated validation loss. `SPA` [177] automatically selects suitable samples for data augmentation. In addition, `Augmix` in [88] is characterized by the combination of multiple basic augmentation procedures paired with a consistency loss. These augmentation methods are chosen at random and stacked, resulting in a rich variety of enhanced images. However, these existing image-level augmentations may create ambiguous images or discard critical information from the original ones, making it difficult for a neural network to learn the shape features or localize important objects. Our `MFG Augment` can better maintain image structures during the augmentation process, as discussed in Sect. 4.3.2.3.

4.2.2 Feature-Level Augmentations

In [54], the authors augment the learned feature space of the training dataset with simple transformations, such as adding noise, interpolating, or extrapolating between them. `Manifold Mixup` [188] extended `Mixup` by applying convex combination to the feature map. `Puzzle Mix` [108] and `FMix` [84] apply saliency maps and Fourier transform to use semantically representative parts of the data when generating features. On the other hand, some learning-based methods are also used to generate features. For example, `FATTEN` [135] utilizes a deep encoder/decoder network architecture to model the feature trajectories. In [152], a variational autoencoder-based feature augmentation method is proposed for acoustic modeling. In [127], `Moment Exchange` is proposed that replaces the moments of the learned features of one training image by those of another, and also interpolates the target labels. It forces the model to extract training signal from the moments in addition to the normalized features. `FeatMatch` [116] learns complex, feature-based transformations as well as incorporates information from class-specific prototypical representations. `SFA` [128] augments feature representation using random noise. In [189], the authors augment features using a generator by playing the GAN minimax game against source features. `TriNet` [30] is an auto-encoder network that is proposed to directly synthesize instance features. In [120], the authors propose a feature augmentation method based on the disentangled representation of intrinsic and bias attributes. `CRAFT` [29] is designed for e-identification tasks that performs cross-view adaptation by automatically measuring camera correlation from cross-view visual data distribution and adaptively conducting feature augmentation to transform the original features into a new adaptive space. Spectral feature augmentation is proposed in [215] for contrastive learning on graphs (and images) where the authors estimate a low-rank approximation for each data in the feature map and subtract that approximation from the map to obtain its complement. `A-FAN` [34] generates adversarial features that integrate flexible scales of perturbation strengths, and the features are re-injected through feature normalization. In [72], the authors propose a hyperbolic feature augmentation method that generates diverse and discriminative features in the hyperbolic space to combat overfitting.

On the other hand, semantic data augmentations are also used to complement traditional augmentation techniques. For example, in [191, 192], the authors propose to translate training samples along many semantic directions in the feature space. In [124], `Attribute Mix` is proposed to mix semantically meaningful attribute features from two images that can significantly improve the recognition performance without increasing the inference budgets. In [213], `ObjectAug` is proposed that decouples the image into individual objects and the background using semantic labels, and each object is augmented individually using traditional augmentation methods, and the augmented objects are finally assembled as augmented image. The authors in [35] propose a reasoning-based implicit semantic data augmentation method that samples new directions from similar categories based on estimated covariance matric for each category. `FASA` [205] generates virtual features on the fly to provide more positive samples for rare classes, and leverages a loss-guided adaptive sampling scheme to avoid over-fitting. Also, `AutoFeature` [138] augments the features following an exploration-exploitation strategy in a reinforcement learning framework. `MetAug` performs a meta-learning technique to augment the features by building the augmentation generator that updates its network parameters by considering the performance of the encoder.

4.3 MFG Augment

Image-level augmentations primarily cater to spatial transformations, providing the model with various visual perspectives and enhancing its robustness to different visual conditions. However, these might not always capture intricate data characteristics essential for some tasks. On the other hand, feature-level augmentations target the abstract representations learned by the model. By augmenting at this level, we ensure that the model is not just focusing on raw visual cues but also on the higher-level features that are crucial for accurate decision-making. By combining both levels of augmentation, we aim to create a comprehensive augmentation strategy that holistically improves model generalization, catering to both visual diversity and abstract feature enhancement. Therefore, in this section, we first give a brief introduction of the MFG theory in Sect. 4.3.1, and then propose and discuss the details of the image-level `MFG augment` and feature-level `MFG augment` in Sects. 4.3.2 and 4.3.3, respectively.

4.3.1 MFG Introduction

To begin with, we first give a brief introduction to the MFG theory. MFG is a mathematical framework that models the interactions of a large number of agents, where each agent's strategy depends on the statistical distribution of the strategies chosen by all other agents. The core idea of an MFG model is that by controlling the group strategy of all agents, the agents' state distribution evolves with time across an optimized path so that their costs that generate along the path can be minimized.

A toy example is that: Imagine a big crowd of people. While each person might decide to move in a certain way, MFG tries to understand how the entire crowd moves based on the average choices people make. This helps in predicting how large groups behave without getting lost in the details of each individual's actions. It's a way to understand big group dynamics without needing to know every tiny detail about everyone. In the proposed `MFG Augment`, the agent is an image's pixel or a learned feature. The agent state is the image's pixel value or the value of its learned feature, and the agent's strategy is a newly introduced variable that is used to change the image's pixel value or feature value.

Mathematically, a typical MFG model consists of a large number of agents denoted by $A = \{a_1, a_2, \ldots, a_k\}$, the agent dynamics is governed by the following stochastic equation:

$$dX^k(t) = f\left(u^k(t)\right) dt + \sigma dW^k(t), \; X^k(0) = x_0^k, \tag{4.1}$$

where $X^k(t)$ is the state of agent a_k, i.e., the "pixel" and "feature" at time t, $u^k(t)$ is the control input, and x_0^k is the initial state of agent a_k at time $t = 0$. f is a function that describes how the agents' control inputs influence their states. $W^k(t)$ is a standard Brownian motion that captures the agent's stochastic property, and σ is its intensity. Under the dynamics constraint, agents aim to minimize the cost in the time interval $[0, T]$ by finding their optimal strategy $u^*(t)$:

$$\min_{u(t)} J_k = \mathbb{E}_X\left[\int_0^T L\left(u(t), X^k(t), \rho(t)\right) dt + G\left(X^k(T), \rho(T)\right)\right], \tag{4.2}$$

where $\rho(t)$ is the distribution of agents' state at time t, known as the mean-field term in MFG theory. Here, $L\left(u(t), X^k(t), \rho(t)\right)$ is known as the running cost, since it is generated continuously in the time interval $[0, T]$, and $G\left(X^k(T), \rho(T)\right)$ is known as terminal cost, since it is generated only at the terminal time T. The agents' optimal control strategy $u^*(t)$ is given by solving the mentioned MFG problem.

Remark 4.1 In the MFG theory, the mean-field term $\rho(t, x)$ serves as a variable that summarizes the state of all the agents. In other words, the relationship and interactions between pixels or the learned features are summarized in $\rho(t)$, which enables us to optimize the data augmentation process from a micro perspective, which is different from existing works that change each pixel or feature independently without considering the relationship between them.

4.3.2 Image-Level MFG Augment

Let $x_k(t) \in \mathbb{R}$, $k = 1, 2, \ldots, N$, be the k-th pixel of an image, where N is the image size. Then, an image can be represented as a distribution over its' pixel value space. **The idea is that:** we transform the distribution of an image's pixels, denoted as $\rho_{Initial}(x)$ into the distribution of another image's pixels $\rho_{Target}(x)$ within a time interval $[0, T]$ along a "path" $\rho^*(t, x)$, so that we can generate new images by sampling from the "path". Figure 4.1 provides a visual illustration of the idea.

4.3.2.1 Problem Formulation of Image-Level MFG Augment

Since the initial point (i.e., when $t = 0$) is the initial image pixel distribution, we have:

$$\rho(0, x) = \rho_{Initial}(x). \tag{4.3}$$

On the other hand, to control the transformation direction of the "path", we define a control variable $u_k(t)$ that can change pixels' values, and the controlling process is described by the following equation:

$$dx_k(t) = u_k(t)dt. \tag{4.4}$$

To summarize the global information of all pixels, we define the mean-field term, i.e., the distribution of the image's pixels as $\rho(t, x)$. Then, according to [117], dynamics in (4.4) is transformed into its distribution dynamics:

$$\partial_t \rho(t, x) + \nabla \cdot (\rho(t, x)u(t, x)) = 0, \tag{4.5}$$

where ∂ is the partial derivative operator, and ∇ is the divergence operator. $u(t)$ is transformed into a field control $u(t, x)$.

We also avoid changing the image too fast along the path $\rho(t, x)$ so that the augmented images in adjacent time slots in the "path" share a certain similarity. Therefore, we quantify the cost function by imposing a penalty on the L2-norm of the control function. The overall MFG augmentation problem at the image level is given as follows:

$$\begin{aligned} &\min_{u,\rho} J = \int_0^T \int_\Omega \rho(t, x) \| u(t, x) \|_2^2 \, dxdt + KL\left(\rho(T, x) \| \rho_{Target}(x)\right) \\ &s.t. \quad \begin{cases} C_1 : \partial_t \rho(t, x) + \nabla \cdot (\rho(t, x)u(t, x)) = 0, \\ C_2 : \rho(0, x) = \rho_{Initial}. \end{cases} \end{aligned} \tag{4.6}$$

where Ω is the pixel value space, i.e., $\Omega = [0, 255]$, of the images. KL is the Kullback–Leibler (KL) divergence that quantifies the distance between the final point in the path $\rho(T, x)$ and the target image's distribution ρ_{Target}, so that the path $\rho(t, x)$ can move towards the target distribution ρ_{Target}. The constraint C_1 in Eq. (4.6) comes from Eq. (4.5) that describes how the augmented images ρ in the "path" are controlled by the control variable u, and the constraint C_2 comes from Eq. (4.3) indicating that the start point of the augmented image path is the initial image's pixel distribution.

4.3.2.2 G-Prox PDHG Algorithm

We utilize the numerical method: G-prox PDHG algorithm, proposed in [98] to solve the proposed image-level `MFG Augment` problem in (4.6). In summary, the algorithm transforms the MFG problem in Eq. (4.6) into a Lagrangian function, and then solves the transformed dual problem. The detailed steps are shown below.

Step 1: Define the Lagrangian function $\mathcal{L}$ based on Eq. (4.6) as:

$$\mathcal{L} = J - \int_0^T \int_\Omega \phi(t, x) \left(\partial_t \rho(t, x) + \nabla \cdot m(t, x)\right) dx dt, \tag{4.7}$$

where $m(t, x) = \rho(t, x)u(t, x)$ is an intermediate variable, $\phi(t, x)$ is the introduced dual variable, and J is the MFG cost function defined in Eq. (4.6).

Therefore, the optimal image-level data augmentation control strategy $u^* = \frac{m^*}{\rho^*}$ is given by optimizing the three optimization variables:

$$\rho^*, m^*, \phi^* = \arg \min_{\rho, m} \max_{\phi} \{\mathcal{L}\} \tag{4.8}$$

Step 2: Update the augmented image path ρ, the intermediate variable m, and the introduced dual variable ϕ iteratively by solving the min-max problem in Eq. (4.8) based on the following iterative strategy.

$$\begin{cases} \rho^{k+1} = \arg \min_{\rho} \left\{ \mathcal{L} + \frac{1}{2\tau} \left\| \rho - \rho^k \right\|_{L^2}^2 \right\}, \\ m^{k+1} = \arg \min_{m} \left\{ \mathcal{L} + \frac{1}{2\tau} \left\| m - m^k \right\|_{L^2}^2 \right\}, \\ \phi^{k+1} = \arg \max_{\phi} \left\{ \mathcal{L}(\bar{\rho}^{k+1}, \bar{m}^{k+1}, \phi^k) - \frac{1}{2\sigma} \left\| \phi - \phi^k \right\|_{H^1}^2 \right\}, \end{cases} \tag{4.9}$$

where τ and σ are small step sizes. $\bar{\rho}^{k+1} = 2\rho^{k+1} - \rho^k$, $\bar{m}^{k+1} = 2m^{k+1} - m^k$, and L^2 and H^1 norm are defined as: $\|u\|_{L^2}^2 = \int_0^T \int_\Omega (u(t, x))^2 dx dt$, $\|u\|_{H^1}^2 = \int_0^T \int_\Omega (\partial_t u(t, x))^2 + \|\nabla u(t, x)\|^2 dx dt$. The summary of the G-prox PDHG algorithm can be found in Algorithm 2.

Algorithm 2 MFG augment in image-level augmentation

1: **Input:** Randomly choose two images with paired labels (x_1, y_1), (x_2, y_2); Two possibility parameters $\alpha, \beta \in [0, 1]$; Two random generated variables $a, b \in [0, 1]$.
2: **if** $a \leq \alpha$ and $b \leq \beta$ **then**
3: **Initialize:** The initial image's pixel distribution $\rho_{Initial}$ by normalizing x_1, and target's image pixel distribution ρ_{Target} by normalizing x_2; Max iterations for G-prox PDHG algorithm K; Step size τ, σ; Terminal Time T.
4: **Define:** the Lagrangian function as in Eq. (4.7).
5: **while** $k \leqslant K$ **do**
6: update ρ, m, ϕ based on Eq. (4.9)
7: $k = k + 1$.
8: **end while**
9: **end if**
10: **Output:** The augmented images $\rho(t_i, x)$ are obtained by sampling the optimal image augmentation path $\rho^*(t, x)$ at time instances $t = t_1, t_2, \ldots, t_N$, where the label of the augmented image $\rho(t_i, x)$ is y_1 if $t_i \leq 0.5T$, and its' label is y_2 if $t_i \geq 0.5T$.

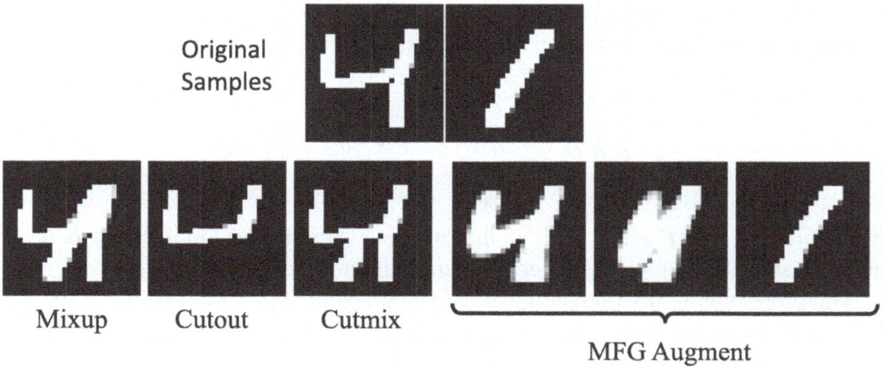

Fig. 4.4 Comparison of generated images with `Mixup`, `Cutout`, `Cutmix`, and `MFG Augment`. Note that `MFG Augment` can keep the edges and shape of objects in the generated images, but `Cutout`,`Cutmix`, and `Mixup` can not

4.3.2.3 What Kind of Images Does Image-Level MFG Augment Generate?

In MFG theory, the value of the control $u(t, x)$ is decided by the time t and pixel x. In other words, at the same time t, as long as two pixels have the same pixel value and the pixels with the same target pixel value, they will have the same control u. This property is beneficial for image augmentation since pixel with the same values usually belongs to one object, and pixels from adjacent locations with different values form an "edge". By using our `MFG Augment`, the generated image can keep the "edge shape" unchanged during the transformation. Figure 4.4 compares `MFG Augment` with `Mixup`, `Cutout`, and `Cutmix`. One can notice that the generated image by `Mixup`, `Cutout`, and `Cutmix` is unnatural and can not keep the "edge" and "shape" as the original samples. However, our `MFG Augment` keeps a great shape with the original samples.

4.3.3 *Feature-Level MFG Augment*

We introduce `MFG Augment` in feature-level augmentation in this section. The idea is inspired by [14, 54, 154], where the authors claim that higher-level representations can expand the relative volume of plausible data points within the feature space. As such, when traversing along the learned manifold, it is more likely to encounter realistic samples in feature space than compared to the input space. Meanwhile, the development of self-supervised learning offers an effective way for learning useful features, e.g., variational autoencoders [110], generative adversarial networks [134], etc.

4.3.3.1 Problem Formulation of Feature-Level MFG Augment

Let $s_k(t) \in \mathbb{R}^m$, $k = 1, 2, \ldots, N$, be the learned features of the k-th image in a dataset, where N is the dataset size and t is time and m is the dimension of the feature. **The idea is that:** The whole dataset is regarded as a distribution over its low-dimensional feature space. We use `MFG Augment` to transform a distribution of a set of images' features that includes N_1 images, denoted as $\rho_{Initial}(s)$, to another distribution of a set of images' features that includes N_2 samples ($N_1 + N_2 = N$), denoted as $\rho_{Target}(s)$, along an optimized "path" $\rho^*(t, s)$ within a time interval $[0, T]$, so that the sampled points in such "path" are all augmented features. Ideally, the optimized "path" is the exact manifold of the images in their latent space. Figure 4.1 describes a conceptive illustration of the idea.

Since the initial point ($t = 0$) and the final point ($t = T$) of the path $\rho(t, s)$ is the distribution of the initial set of features and the distribution of the target set of features, we have:

$$\rho(0, s) = \rho_{Initial}(s), \tag{4.10}$$

$$\rho(T, s) = \rho_{Target}(s). \tag{4.11}$$

On the other hand, to control the transformation direction of the "path", we also define a control variable $u(t)$ that can control the flow of the distribution of features. The controlling process is described by,

$$ds_k(t) = u_k(t)dt. \tag{4.12}$$

Similar to Eq. (4.5), in order to summarize the global information of all learned features in the feature space, we again define the mean-field term, i,e,m the distribution of the learned features as $\rho(t, s)$. According to [117], the state-control dynamics of the augmented feature distribution is given by:

$$\partial_t \rho(t, s) + \nabla \cdot (\rho(t, s)u(t, s)) = 0. \tag{4.13}$$

Similar to image-level augmentation, we aim to generate a "smooth" path between two distributions. We also impose an L-2 norm penalty on the control in the cost function. Therefore, the overall feature-level `MFG Augment` problem is given as:

$$\min_{u,\rho} J = \int_0^T \int_\Omega \rho(t,s)\, \|u(t,s)\|_2^2 \, ds dt \tag{4.14}$$

$$s.t. \quad \begin{cases} C_1 : \partial_t \rho(t,s) + \nabla \cdot (\rho(t,s) u(t,s)) = 0, \\ C_2 : \rho(0,s) = \rho_{Initial}, \\ C_3 : \rho(T,s) = \rho_{Target}, \end{cases} \tag{4.15}$$

where Ω is the image's feature space (the latent space). The constraint C_1 comes from Eq. (4.13) which describes how the augmented feature distribution ρ is controlled by the introduced control variable u. The constraint C_2 and C_3 come from Eqs. (4.10) and (4.11) indicating that the start point and the end point of the augmented feature path is the distribution of the initial set of features and the distribution of the target set of features.

4.3.3.2 APAC-Net Network to Solve Feature-Level MFG Augmentation Problem

The MFG problem in Eq. (4.14) is a high-dimensional problem because the learned features are usually hundreds or thousands-dimensional. According to [12], the grid-based numerical methods, such as PDHG and Adjoint method [167], are prone to the curse of dimensionality, i.e., their computational complexity grows exponentially with the spatial dimension. Thus, we utilize the APAC-Net, an alternating population and agent control neural network approach geared toward high-dimensional MFG problems [132] to solve it. In particular, we solve Eq. (4.14) by training a Wasserstein Generative Adversarial Network (WGAN)-like network. A detailed description of the implementation of the APAC-Net is shown below.

First, we transform the MFG problem into the Lagrangian problem:

$$\min_{\rho,u} \max_{\phi} \{\mathcal{L}\} = \int_0^T \int_\Omega \rho(t,s)\, \|u(t,s)\|_2^2 - \phi(t,x)\, (\partial_t \rho(t,s) + \nabla \cdot (\rho(t,s) u(t,s)))\, ds dt. \tag{4.16}$$

Second, we solve (4.16) by training a WGAN-like neural network named APAC-Net.

The neural networks generator is denoted by $G_\theta(s,t)$, and the discriminator is denoted by $N_\omega(s,t)$. The generator generates the path (i.e., the manifold), and the discriminator judges whether the generated path satisfies the optimality condition. Then we set

$$\phi_\omega(z,t) = (1-t)N_\omega(z,t), \tag{4.17}$$

$$G_\theta(s,t) = (1-t)s_0 + t(1-t)N_\theta(s,t) + ts_1, \tag{4.18}$$

where $s_0 \sim \rho_{Initial}$ are samples drawn from the initial distribution $\rho_{Initial}$, and $s_1 \sim \rho_{Target}$ are samples drawn from the target distribution. Note that the formulation of $G_\theta(s,t)$ automatically encodes the boundary conditions in Eqs. (4.10) and (4.11).

During the training process of the APAC-Net, we alternatively train $G_\theta(s,t)$ and $\phi_\omega(s,t)$. Specifically, we first sample a batch $\{s_0^b\}_{b=1}^B$ from the initial distribution, a batch $\{s_1^b\}_{b=1}^B$ from the target distribution ρ_{Target}, and $\{t^b\}_{b=1}^B$ uniformly from the time interval $[0,T]$. Then we compute the output of the generator $z_b = G_\theta(s,t)$ as in Eq. (4.18). The total loss of the discriminator ϕ_ω is then given as the value function:

$$l_\phi = \frac{1}{B}\sum_{b=1}^{B} \phi_\omega(z_b, 0) + \partial_t\phi_\omega(z_b, t_b) - |\nabla_z\phi_\omega(z_b, t_b)|^2, \tag{4.19}$$

where we can optionally add a regularization term that gives the optimality Karush–Kuhn–Tucker (KKT) condition of Eq. (4.14), which is also known as the Hamilton-Jacobi-Bellman (HJB) equation in MFG theory:

$$l_{HJB} = \lambda\frac{1}{B}\sum_{b=1}^{B} \left\| \partial_t\phi_\omega(z_b, t_b) - |\nabla_z\phi_\omega(z_b, t_b)|^2 \right\|, \tag{4.20}$$

to penalize the derivations from the HJB equation. Finally, we back-propagate the total loss to update the weights of the discriminator ϕ_ω. To train the generator, we again sample $\{s_0^b\}_{b=1}^B$, $\{s_1^b\}_{b=1}^B$, and $\{t^b\}_{b=1}^B$ as before, and compute the loss of the generator:

$$l_G = \frac{1}{B}\sum_{b=1}^{B} \partial_t\phi_\omega\left(G_\theta(s_b, t_b), t_b\right) - |\nabla_z\phi_\omega(z_b, t_b)|^2. \tag{4.21}$$

At last, we back-propagate the loss to the weights of G_θ.

Once the optimal path $\rho^*(t,s)$ (i.e, the manifold) is generated, augmented features can be obtained by sampling $\rho^*(t,s)$ at any time $t \in [0,T]$. The augmented features can be used as input for a learning task or as the basis for generating new images. The training process of the APAC-Net is described in Algorithm 3.

Algorithm 3 MFG augment in feature-level augmentation

Input: The neural networks G_θ and ϕ_ω; Two sets of features with paired labels $(s_i^n, y_i^n)_{i=0,1}$; Two possibility parameters $\alpha, \beta \in [0, 1]$; Two random generated variables $a, b \in [0, 1]$.

if $a \leq \alpha$ and $b > \beta$ **then**

  Initialize: G_θ and ϕ_ω as in (4.17) and (4.18); The initial distribution of features $\rho_{Initial}$ by scaling s_0^n; The target distribution of features ρ_{Target} by scaling s_1^n.

  while not converge **do**

    Train ϕ_ω for N epochs:

    Sample batch (s_0^b, t_b), (s_1^b, t_b) where $s_0^b \sim \rho_{Initial}$, $s_1^b \sim \rho_{Target}$, and $t^b \sim \text{Unif}(0, T)$.

    $z_b \leftarrow G_\theta(s_b, t_b)$ for $b = 1, 2, \ldots, B$

    $l_\phi \leftarrow \frac{1}{B} \sum_{b=1}^{B} \phi_\omega(z_b, 0) + \partial_t \phi_\omega(z_b, t_b) - |\nabla_z \phi_\omega(z_b, t_b)|^2$

    $l_{HJB} \leftarrow \lambda \frac{1}{B} \sum_{b=1}^{B} \left\| \partial_t \phi_\omega(z_b, t_b) - |\nabla_z \phi_\omega(z_b, t_b)|^2 \right\|$

    Back-propagate the loss $l_{total} = l_\phi + l_{HJB}$ to weights ω.

    Train G_θ M epochs:

    Sample batch (s_0^b, t_b), (s_1^b, t_b) where $s_0^b \sim \rho_{Initial}$, $s_1^b \sim \rho_{Target}$, and $t^b \sim \text{Unif}(0, T)$

    $l_G \leftarrow \frac{1}{B} \sum_{b=1}^{B} \partial_t \phi_\omega \left(G_\theta(s_b, t_b), t_b\right) - |\nabla_z \phi_\omega(z_b, t_b)|^2$

    Back-propagate the loss l_G to weights θ.

  end while

end if

Output: The augmented features set $\rho(t_i, s)$ are obtained by sampling the optimal path $\rho^*(t, s)$ at time instances $t = t_1, t_2, \ldots, t_N$, where the label of the augmented features set $\rho(t_i, s)$ is y_1 if $t_i \leq 0.5T$, and its' label is y_2 if $t_i \geq 0.5T$.

Remark 4.2 In the image-level MFG augmentation problem in Eq. (4.6), we use the G-prox PDHG algorithm to solve it because as a numerical method, it can guarantee the convergence and optimality of the solution. However, the G-prox PDHG algorithm does not scale well to high-dimensional MFG problems as the feature-level optimization problem in Eq. (4.14), so here we train the APAC-Net network to solve it. On the other hand, although APAC-Net is able to solve high-dimensional MFG problems, it can not guarantee the convergence and optimality of its solution. Therefore, for the low-dimensional image-level MFG augmentation problem in Eq. (4.6), we use the G-prox PDHG algorithm to solve it.

4.4 Theoretical and Performance Analysis of MFG Augment

In Sect. 4.4.1, we first analyze why `MFG Augment` generates "High-Quality" data by introducing the physical meaning of the proposed data augmentation optimization problem in Eqs. (4.6) and (4.14). Then we propose a quantifiable metric for the proposed data augmentation process in Sect. 4.4.2, and give the complexity analysis in Sect. 4.4.3, respectively.

4.4.1 Analysis of the Proposed MFG Data Augmentation Optimization Problem

In Eqs. (4.6) and (4.14), the physical meaning of the term $\int_0^T \int_\Omega \rho(t,x)\|u(t,x)\|_2^2 dxdt$ is the Wasserstein-2 (W-2) distance between the generated data distribution and the source data distribution, and we aim to minimize such W-2 distance. In existing works, the idea of minimizing the distance between generated data distribution and source distribution has been widely utilized in the design of Generative Adversarial Network (GAN) [81]. However, there exists the issue of Perfect Discriminator and Unstable Training since the loss function of GAN is a variation of Kullback–Leibler (KL) divergence, which is meaningless when two distributions have no overlap (i.e., log(0) is meaningless), and thus results in gradient vanishing [193]. To deal with the hard training of GAN, Wasserstein GAN [8] is proposed that utilize the Wassterin distance in the loss function instead of KL divergence. Compared with KL divergence, Wasserstein distance is symmetric and is able to quantify the distance between two distributions even if they have no overlap [4]. Therefore, in the proposed `MFG Augment`, the utilization of W-2 distance in the MFG framework enables us to generate data that has the smallest W-2 distance with source data (if the solution that we obtain is optimal). In addition, the utilization of MFG theory allows us to quantify the "quality" of the generated data distribution using the pre-defined Hamilton-Jacobi-Bellman (HJB) and the Fokker-Planck-Kolmogorov (FPK) equation, which is introduced in Sect. 4.4.2.

4.4.2 Quantifiable Optimality Condition of the Proposed MFG Augment Scheme

In the proposed `MFG Augment`, we formulate the data augmentation process as an MFG problem, and such formulation enables us to evaluate the performance of such data augmentation process. Specifically, in order to calculate the optimal data augmentation strategy $u(t,s)$ and $u(t,x)$ in Eqs. (4.6) and (4.14), the following partial differential equation (PDE) should be satisfied:

$$\partial_t \rho - \frac{\sigma^2}{2}\Delta\rho - \nabla \cdot \left(\rho \nabla_p H\left(x, \rho, \nabla\phi\right)\right) = 0, \tag{4.22}$$

$$-\phi_t - \frac{\sigma^2}{2}\Delta\phi + H(x, \rho, \nabla\phi) = 0, \tag{4.23}$$

$$\rho(0,x) = \rho_0(x),\ \phi(T,x) = G\left(X(T), \rho(T,x)\right), \tag{4.24}$$

where Eq. (4.22) is named the Fokker-Planck-Kolmogorov (FPK) equation as introduced in Eq. (4.5). Equation (4.23) is called the Hamilton-Jacobi-Bellman

(HJB) equation that gives the optimality condition of the data augmentation process. Equation (4.24) is the boundary condition of the PDE system.

Traditional metrics often evaluate data augmentation techniques based on heuristic measures such as test accuracy, which might not fully capture the specific optimality conditions that the proposed MFG Augment. However, given the theoretical foundation of our method in MFG, it is essential to have a metric that is compatible with its underlying principles. Here, we utilize the FPK and the HJB equation as a metric to evaluate the performance of the proposed data augmentation process, and the introduction of such quantifiable metrics is driven by the distinct nature of our proposed data augmentation method. Empirically, we give the residual of the FPK and the HJB equation using image-level `MFG Augment` on Mnist in Fig. 4.5a using Algorithm 2. It can be observed that after about 1000 iterations the residuals become lower than 0.0001. Similarly, in Fig. 4.5b, the residual of the HJB equation is depicted using feature-level `MFG Augment` on CIFAR-10. It is shown that after about 10,000 training epochs the HJB residual becomes lower than 0.0001, which serves as a quantifiable metric to evaluate the performance of the data augmentation process besides model accuracy.

4.4.3 Complexity Analysis

In Algorithm 2, to find the optimal data augmentation strategy $u(t, x)$, the variables: $\rho(t, x)$, $m(t, x)$, and $\phi(t, x)$ are updated for each time step and pixels. Therefore, the computational complexity of the image-level `MFG Augment` is $O(N \times D)$, where N is the number of time discretization slots, and D is the image size. Empirically, the time utilized to obtain optimal data augmentation strategies for image-level `MFG Augment` is 20 $\sim$ 40 seconds between every two images for Mnist, 40 $\sim$ 60 seconds for CIFAR-10, and 8 $\sim$ 9 mins for ImageNet. On the other hand, in Algorithm 3, we train a GAN-like network to find the feature-level augmentation strategy. Empirically, the time utilized to obtain optimal data augmentation strategies for feature-level `MFG Augment` is: 1.4 $\sim$ 1.6 hours for MNIST, 1.8 $\sim$ 2.0 hours for CIFAR-10, and 3.0 $\sim$ 3.4 hours for ImageNet, respectively. The experiments are implemented using CPU i5-12600 and GPU NVIDIA K80.

It is worthwhile to mention that image-level `MFG Augment` is designed to generate augmented images between every two original images, while feature-level `MFG Augment` is used to generate augmented features between every two random sets of learned features. In other words, Algorithm 2 needs to be repeated between every two original images, while the GAN in Algorithm 3 is trained once on the whole original dataset distribution, which makes the image-level `MFG Augment` is more suitable for small datasets and feature-level `MFG Augment` are more efficient for large datasets.

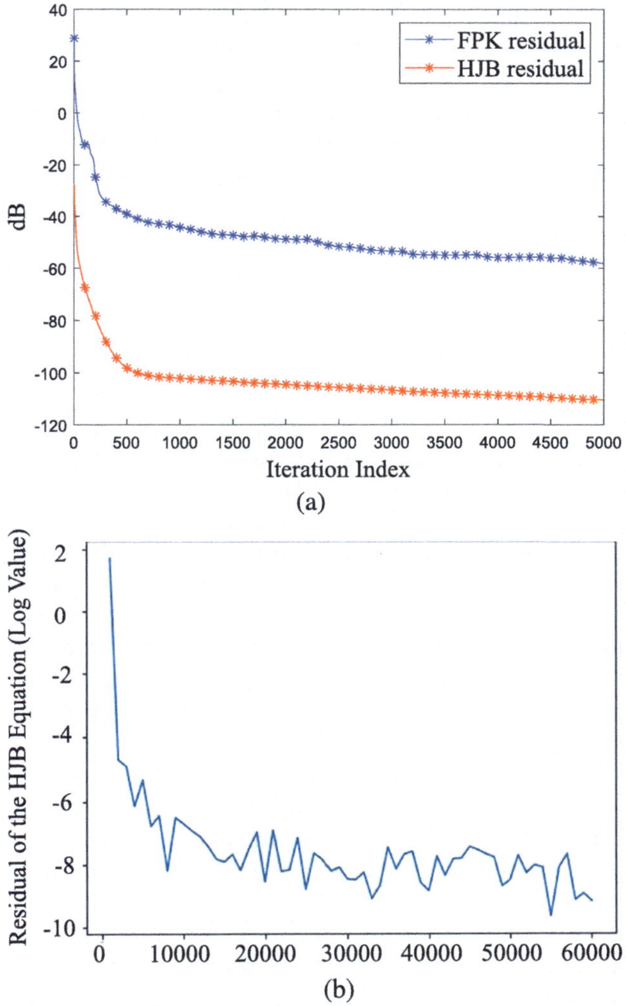

Fig. 4.5 Residuals of the FPK and HJB equation in `MFG Augment`. (**a**) Residual of the FPK and the HJB equation in image-level Augmentation, implemented on Mnist. (**b**) Residual of the HJB equation in image-level Augmentation, implemented on CIFAR-10

4.5 Experiments

In this section, we empirically evaluate our data augmentation method `MFG Augment`, and compare it with existing state-of-art data augmentation methods including `Cutout` [55], `Cutmix` [204], and `Augmix` [88]. These methods are widely recognized and established in the domain, offering diverse augmentation paradigms from occlusion-based to region-mixing and complex blending of augmentations. By comparing against these recognized techniques, our evaluation gains rigor and relevance. We implement the augmentation methods on datasets: MNIST [51], CIFAR-10 [114], and ImageNet [52] on several neural network architectures including EfficientNet [178], ResNet-18 [85], and Resnet-50 [85]. All the experiments are carried 5 times, and the average performance is reported. In summary, we observe significant improvements in our `MFG Augment` over the three comparative augmentation methods on both MNIST and CIFAR-10 for all tested neural network architectures.

4.5.1 MFG Augment Implementation Settings

For the implementation of our `MFG Augment`, we combine both the image-level augmentation and the feature-level augmentation. We combine the augmented images, decoded augmented features, and the original dataset into the new trainset. Specifically, the two possibility parameters in Algorithms 2 and 3 are set as $\alpha = 0.4$ and $\beta = 0.2$, respectively. That is, given two random data, the possibility that we implement our `MFG Augment` is 0.4, and the possibility that we implement image-level augmentation and feature-level augmentation is 0.2 and 0.8, respectively. The probability for data augmentation $\alpha = 0.4$, is determined through preliminary tests that balance data diversity with the original data's integrity. Also, we allocate different probabilities to image-level versus feature-level augmentations (i.e., $\beta = 0.2$), which is designed to maximize our method's effectiveness. In addition, we set the time interval $[0, T]$ to be $[0\,\text{s}, 1\,\text{s}]$ for both image-level and feature-level implementation, and we generate the augmented images and features by sampling 2 time-instances for every generated path $\rho^*(t, x)$ at time instants $t = 0.23\,\text{s}, 0.77\,\text{s}$. (Theoretically, we can sample an infinite number of augmented data since $\rho^*(t, x)$ is time-continuous. Here, the choice of the two time-instants at $t = 0.23\,\text{s},\ 0.77\,\text{s}$ is determined by a series of preliminary test, and such settings achieves the optimal performance for our proposed `MFG Augment` method). For the step sizes in Algorithm 2, we set $\tau = 9.9,\ \sigma = 0.1$, and the max iteration is set to be $K = 3,000$. For the training of the APAC-Net in Algorithm 3, we use a ADAM optimizer with $\beta = (0.5, 0.9)$, learning rate 0.0005 for ϕ_ω, learning rate 0.0001 for G_θ, weight decay of 0.0001 for both ϕ_ω and G_θ.

4.5.2 MNIST

4.5.2.1 Cutout, Cutmix, Augmix Implementation Settings

For a fair and clear comparison, we do not use other data augmentation settings such as random-cropping, or flipping. For `Cutout`, the paste-back size is set to be 8. For `Cutmix`, the hyperparameter: cutmix probability, is set to be 0.5, and the beta distribution $\beta(\alpha, \alpha)$, is set to be $\alpha = 1.0$, as suggested in [204]. For Augmix, the augmentation severity is set to be 2. Other augmentation hyperparameters follow the settings in [55, 88, 204].

4.5.2.2 Training Settings

We first implement augmentations on the state-of-art neural network architecture: EfficientNet [178]. We optimize it with stochastic gradient descent using Nesterov momentum and train it for 100 epochs. The initial learning rate is set to be 0.01 [139]. Following [88, 204], we use a weight decay of 0.0001 for `Cutmix` and `Augmix`, and 0.0005 otherwise. Other hyperparameters follow the settings in [178].

We first train the EfficientNet with several reduced MNIST datasets with a variety of trainset samples from 50 to 2000, and compare the test accuracy of the proposed `MFG Augment` with `Cutout`, `Cutmix`, and `Augmix` on these reduced MNIST trainsets.

4.5.2.3 Results

As is shown in Table 4.1, we observe that our `MFG Augment` consistently and significantly improve over `Cutout`, `Cutmix`, and `Augmix` with all reduced trainsets. Specifically, when the data size of the reduced MNIST trainset is 50, our MFG Augment achieves the test accuracy gain $+$ 59.26% compared with the baseline and even achieves the accuracy gain at +11.2%, +13.2%, and +3.6% over `Cutout`, `Cutmix`, and `Augmix`, respectively. In addition, we can achieve a test accuracy of 98.52% even by training a reduced MNIST trainset with only 2000 samples, while the standard MNIST needs to train all the 60,000 samples to get a comparable test accuracy.

We also compare the test accuracy on other state-of-art neural network architectures: ResNet-18 [85], and ResNet-50 [85]. All the hyperparameter settings are the

Table 4.1 Test accuracy (%) on reduced MNIST with trainset sizes varying from 50–2000, trained with EfficientNet. Note that the full MNIST has 60,000 samples

Reduced trainset size	Baseline	Cutout	Cutmix	Augmix	MFG Augment
50	11.35	59.41	57.18	67.01	**70.61**
100	31.23	65.95	68.53	77.68	**80.26**
200	64.83	73.67	78.35	86.8	**88.94**
400	75.33	87.13	84.59	92.34	**94.95**
600	82.80	90.92	90.61	93.9	**96.34**
800	91.40	93.16	92.02	95.95	**97.96**
2,000	94.32	96.03	96.18	97.35	**98.52**

The bolded values represent the highest accuracy in each row

Table 4.2 Test accuracy (%) on reduced MNIST, trainset size = 2000. Note that the original MNIST has 60,000 samples

Model	EfficientNet	ResNet-18	ResNet-50
Cutout	96.03	97.66	96.98
Cutmix	96.18	96.77	96.40
Augmix	97.35	98.50	98.16
MFG Augment	**98.52**	**99.02**	**98.62**

The bolded values represent the highest accuracy in each column

same as in EfficientNet, except we train both networks for 150 epochs to guarantee convergence. The trainset is still the reduced MNIST with a data size of 2000. As can be observed in Table 4.2, our `MFG Augment` achieves the highest test accuracy over all three comparative augmentation methods. Specifically, we achieve the test accuracy 99.02% on ResNet-18, which is +1.36% better than `Cutout`, +2.25% better than `Cutmix`, and +0.52% better than `Augmix`. Also, on EfficientNet and ResNet-50, we achieve state-of-art test accuracy of 98.52 and 98.62% on reduced MNIST (trainset size=2000), respectively.

4.5.3 CIFAR-10

4.5.3.1 Cutout, Cutmix, and Augmix Implementation Settings

For a fair and clear comparison, we do not use other data augmentation settings such as random cropping or flipping. For `Cutout`, we set the cutout paste-back size to be 16, as suggested by DeVries and Taylor [55]. For `Cutmix`, the hyperparameter: cutmix probability, is set to be 0.5, and the beta distribution $\beta(\alpha, \alpha)$, is set to be 1.0, as suggested in [204]. For `Augmix`, the augmentation severity is set to be 3. All the other augmentation hyperparameters follow the settings in [55, 88, 204].

4.5.3.2 Training Settings

Similar to the above experiments, we first train the EfficientNet with several reduced CIFAR-10 trainsets with trainset sizes varying from 100 to 50,000. The training settings, such as momentum, scheduler, etc., are the same as in the MNIST experiment, except the initial learning rate is 0.1, and we train the neural network for 250 epochs for convergence.

Table 4.3 Test accuracy (%) on CIFAR-10 using several reduced trainset with datasize vary form 100–50,000, trained with EfficientNet

Reduced trainset size	Baseline	Cutout	Cutmix	Augmix	MFG augment
100	17.26	18.10	20.25	21.20	**29.21**
400	26.78	30.52	30.55	32.93	**52.51**
800	31.33	31.43	33.97	37.50	**61.14**
2,000	41.16	41.01	43.60	48.55	**65.11**
10,000	69.89	71.44	69.49	74.99	**83.23**
20,000	79.5	79.23	79.76	82.77	**86.87**
30,000	83.33	83.73	84.42	85.95	**88.50**
40,000	85.09	85.15	87.17	87.93	**89.95**
50,000	86.46	88.13	88.94	88.95	**90.43**

The bolded values represent the highest accuracy in each row

Table 4.4 Test accuracy (%) on CIFAR-10

Model	EfficientNet	ResNet-18	ResNet-50
Cutout	88.13	88.27	89.61
Cutmix	88.94	91.37	92.49
Augmix	88.95	91.34	91.95
MFG Augment	**90.43**	**94.76**	**94.69**

The bolded values represent the highest accuracy in each column

4.5.3.3 Results

As shown in Table 4.3, we again observe consistent improvements in our `MFG Augment` over the three comparative augmentation methods. Especially, our `MFG Augment` gains impressive results when the dataset size is small.

When the trainset size is 2,000, our `MFG Augment` achieves 65.11% test accuracy, beating the `Cutout` 41.01%, Cutmix 43.60% and Augmix 48.55%. We again observe significant improvements when trained with all the 50,000 training samples. Specifically, our `MFG Augment` can achieve an accuracy of 90.43%, which is +3.97% higher than the baseline, and +2.30% higher than `Cutout`, +1.49% higher than `Cutmix`, and +1.48% higher than `Augmix`.

Again, we implement the augmentation methods on ResNet-18 and ResNet-50. We train with all the 50,000 training samples in CIFAR-10. The hyperparameter settings are the same as the previous EfficientNet of CIFAR-10, except we train the networks for 300 epochs for convergence. The results are shown in Table 4.4. Our `MFG Augment` significantly outperforms three comparative augmentation methods on all the tested neural network architectures. Specifically, on ResNet-18, `MFG Augment` achieves 94.76% test accuracy, which is +6.49% higher than `Cutout`, +3.39% higher than `Cutmix`, and +3.42% higher than `Augmix`. On EfficientNet and ResNet-50, our `MFG Augment` again achieves the state-of-art test accuracy at 90.43% and 94.69%, respectively.

4.5.4 ImageNet

4.5.4.1 Cutout, Cutmix, and Augmix Implementation Settings

All images are first pre-processed with standard random cropping horizontal mirroring. For `Cutout`, the mask size is set to 112×112, and the location for dropping out is uniformly sampled. For `Cutmix`, the cutmix probability is set as 1.0, and the beta distribution $\beta(\alpha, \alpha)$ is set to be 1.0, as suggested in [204]. For `Augmix`, the augmentation severity is set to 1.0, and the depth and width of augmentation chains are set to -1 and 3, as suggested in [88].

4.5.4.2 Training Settings

We first train several reduced ImageNet trainsets with ResNet-18, and we follow the standard scheme of [85]. Then, we train the full ImageNet with Resnet-50 scheme of [85]. We found that `Cutout` and `Cutmix` require many training epochs to converge. Therefore, we optimize all the models using 300 epochs with an initial learning rate 0.1, decaying by 0.1 every 75 epochs. The batch size is 256.

4.5.4.3 Results

As is shown in Table 4.5, `MFG Augment` substantially improves the performance over the three baseline augmentation methods. The improvement is especially prominent when training on small ImageNet datasets. When training on 1/4 Imagenet, our `MFG Augment` can achieve test accuracy at 61.13%, which is +6.06% higher than `Cutout`, +3.56% higher than `Cutmix`, and +4.1% higher than `Augmix`. In addition, When training on full Imagenet, our `MFG Augment` can achieve the highest test accuracy at 72.26%.

When trained on ResNet-50, our `MFG` Augment again achieves the highest test accuracy. As is shown in Table 4.6, on ResNet-18, our `MFG Augment` achieves 72.26% test accuracy, which is +2.01% higher than `Cutout`, +0.74% higher than `Cutmix`, +0.77% higher than `Augmix`. On ResNet-50, our `MFG Augment` achieves

Table 4.5 Top-1 accuracy (%) on several reduced ImageNet, trained with ResNet-18

Reduced trainset size	Baseline	Cutout	Cutmix	Augmix	MFG augment
1/4 ImageNet	54.18	55.07	57.57	57.03	**61.13**
1/2 ImageNet	60.92	61.39	61.97	62.77	**65.22**
3/4 ImageNet	65.75	67.04	69.65	68.91	**69.73**
Full ImageNet	69.76	70.25	71.52	71.49	**72.26**

The bolded values represent the highest accuracy in each row

Table 4.6 Top-1 test accuracy (%) on ImageNet

Model	ResNet-18	ResNet-50
Baseline	69.76	76.15
Cutout	70.25	77.01
Cutmix	71.52	78.41
Augmix	71.49	77.68
MFG augment	**72.26**	**79.22**

The bolded values represent the highest accuracy in each column

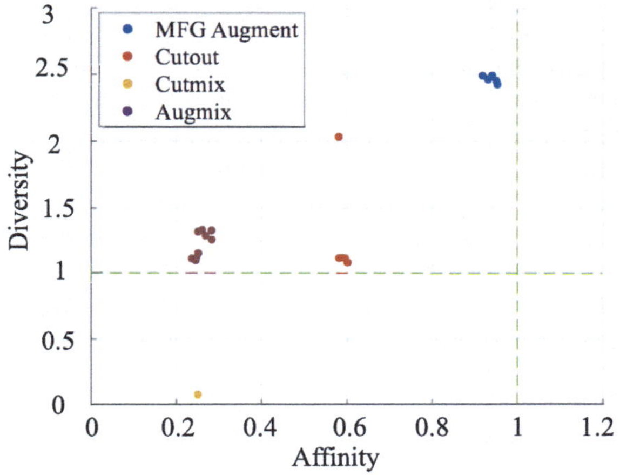

Fig. 4.6 Affinity and diversity plane for Cutout, Cutmix, Augmix, and our MFG Augment on CIFAR-10, trained with EfficientNet

79.22% test accuracy, which is +2.21% higher than `Cutout`, +0.81% higher than `Cutmix`, +1.54% higher than `Augmix`.

4.5.5 *Affinity and Diversity Analysis*

In this subsection, we compare our `MFG Augment` with the state-of-art augmentation methods: `Cutout`,`Cutmix` and `Augmix`, in terms of "Affinity" and "Diversity", proposed in [79]. The experiment is trained with EfficientNet on CIFAR-10 database. All the hyperparameter settings are the same as in the experiments of CIFAR-10 in Sect. 4.5.3. According to [79], test accuracy usually improves when moving to the upper right region in the "Affinity-Diversity plane". We can observe in Fig. 4.6 that our `MFG Augment` is in the upper right corner of this plane, which means that it has the highest affinity and diversity among all the comparative augmentation methods. Specifically, the affinity of our `Augment` is about 0.92, and the diversity is about 2.5, while for `Cutout`, `Cutmix` and `Augmix`, all have affinity < 0.72 and diversity < 2.1. There is an outlier for `Cutout` and `Cutmix`, respectively. The results serve as an explanation for why `MFG Augment` is more effective than `Cutout`, `Cutmix` and `Augmix`.

4.6 Chapter Summary

In this chapter, we propose `MFG Augment`, a data augmentation based on the Mean-field Game theory that can generate a "path" in images' pixel space and feature space, and the points along the path are augmented images' pixels or features. MFGs introduce a systematic and theoretically grounded approach when applied to data augmentation. By representing the collective behavior of individual agents (i.e, image pixels and learned features in the proposed `MFG Augment`), it provides a mathematical framework to generate and analyze new data that's consistent, relevant, and theoretically validated. We demonstrate that our `MFG Augment` achieves state-of-the-art test accuracy on MNIST, CIFAR-10, and ImageNet. The high performance is explained by the increased diversity and affinity of the augmented data. Importantly, `MFG Augment` generates impressive results for small datasets. However, the time complexity of the proposed `MFG Augment` can be high for large datasets. Also, another concern is ensuring the augmented data remains true to real-world scenarios. Looking ahead, there is potential to explore how to reduce the computation complexity in `MFG Augment` for larger datasets. Additionally, combining `MFG Augment` with existing data augmentation techniques to further improve the model performance may also be considered in our future work.

Chapter 5
Mean Field Game Guided Deep Reinforcement Learning

In this chapter, we propose a MFG-guided deep deterministic policy gradient (DDPG) for the task placement in the cooperative MEC, which can help servers make timely task placement decisions, and significantly reduce average service delay. Instead of applying MFG or DRL separately, we jointly leverage MFG and DRL for task placement, and let the equilibrium of MFG guide the learning directions of DRL. We also ensure that the MFG and DRL approaches are consistent with the same goal. Specifically, we novelly define a mean field guided Q-value (MFG-Q), which is an estimation of Q-value with the Nash equilibrium gained by MFG. We evaluate the proposed method's performance using real-world user distribution. Through extensive simulations, we show that the proposed scheme is effective in making timely decisions and reducing the average service delay. Besides, the convergence rates of our proposed method outperform the pure DRL based approaches.

5.1 Task Placement System Model

Consider a basic cooperative MEC system which consists of multiple end users and multiple MEC servers, and all of the end users (e.g. smartphones, laptops and Internet of Vehicles) are covered by MEC servers (e.g. small base stations and wireless access points) at their current locations. End users have their tasks to deal with, and they will upload the tasks to the MEC server for executing over wireless communication. After completing the tasks, the servers will return the results of the tasks to the end users. Once a user's task is uploaded, the task may be placed to be executed on the current server or an adjacent server by the guidance of the cloud server, which means the task can be transferred among the MEC servers.

We consider a scenario consisting of the number of $\mathcal{D}(j \in \mathcal{D} = 1, 2, \ldots, D)$ MEC servers and the number of $\mathcal{U}(i \in \mathcal{U} = 1, 2, \ldots, U)$ end users covered

Y. Kang et al., *Mean Field Guided Machine Learning*, Wireless Networks,
https://doi.org/10.1007/978-3-031-91859-9_5

by servers. Each server is modeled by $M/M/1$ queue. We assume that each user has only one task, so the index of the user equals the index of the task. Different users can run different applications, and different applications may contain different amounts of computations. Furthermore, each task can only be run on one server at one time. In this case, the computational generation rate for user i is defined as λ_i, which means that user i generates λ_i computation units (CPU cycles) per second. Assume that all of the users are generating the number of computations according to a Poisson process with the same mean rate $\bar{\lambda}$. Thus, if the task of user i is executed in MEC server j, the expected service delay in server j can be defined as

$$t_j = \frac{1}{\mu_j - \sum_{i=1}^{U} x_{i,j}\lambda_i}, \tag{5.1}$$

where μ_j is the service rate of MEC server j, $x_{i,j}$ is defined to describe the connection situation between user i and server j. If the task of user i is executed in server j, $x_{i,j} = 1$; otherwise, it is set to be 0. The time consumption t_{ij} contains both the waiting and execution delay. The denominator is the stable processing speed (the amount of computations executed per second), i.e., the frequency for server j. We can find that the processing speed (frequency) decreases as the tasks load increasing in the server.

Accordingly, the expected energy consumption for the user's task executed in server j is

$$e_j = \kappa_j(\mu_j)^2, \tag{5.2}$$

where κ_j is defined as the effective capacitance coefficient that depends on the chip architecture at the MEC server. Due to the different computational capacity and CPU architectures of different MEC servers, service rate μ and effective capacitance coefficient κ are different for each MEC server.

Our goal is to reduce the service delay of the entire system by properly placing the servers that handle users' tasks. In the meantime, we don't want to sacrifice too much energy while reducing latency. Therefore, the balance factor ζ is introduced, and the cost function of users in each server can be defined as

$$l_j = t_j + \zeta e_j. \tag{5.3}$$

ζ is a factor to balance the two components in the cost function, and e_j can be considered as a penalty. Thus, we have the cost function of the system, which can be defined as the expected service delay and energy consumption per user. We try to reduce the average service delay and don't want to increase too much energy consumption during the process. Therefore, the cost function can be described as

$$
\begin{aligned}
L =& \frac{1}{U} \sum_{j=1}^{D} \sum_{i=1}^{U} x_{i,j} \lambda_i f_j \\
=& \frac{1}{U} \sum_{j=1}^{D} \sum_{i=1}^{U} \frac{x_{i,j}\lambda_i}{\mu_j - \sum_{i=1}^{U} x_{i,j}\lambda_i} + \frac{1}{U} \zeta \sum_{j=1}^{D} \sum_{i=1}^{U} \kappa_j x_{i,j} \lambda_i (\mu_j)^2 \\
=& \frac{1}{U} \sum_{j=1}^{D} \frac{N_j \bar{\lambda}}{\mu_j - N_j \bar{\lambda}} + \frac{1}{U} \zeta \sum_{j=1}^{D} \kappa_j N_j \bar{\lambda} (\mu_j)^2 \\
=& \sum_{j=1}^{D} s_j \frac{\bar{\lambda}}{\mu_j - N_j \bar{\lambda}} + \zeta \sum_{j=1}^{D} s_j \kappa_j \bar{\lambda} (\mu_j)^2.
\end{aligned}
\tag{5.4}
$$

U is the number of end users in the system, N_j is the number of users whose tasks are executed in server j. s_j is the proportion of tasks in server j, which is equal to N_j/U. Our object is to minimize the cost function L by scheduling the user's pending task on different servers with the following constrains. The minimization problem can be formulated as

$$
\begin{aligned}
\min_{\mathbf{s}} \ L =& \sum_{j=1}^{D} s_j \frac{\bar{\lambda}}{\mu_j - s_j U \bar{\lambda}} + \zeta \sum_{j=1}^{D} s_j \kappa_j \bar{\lambda} (\mu_j)^2 \\
s.t. \quad & \lambda_i > 0, \forall i \in U \\
& \sum_{i=1}^{U} x_{i,j} \lambda_i < \mu_j, \forall j \in D.
\end{aligned}
\tag{5.5}
$$

5.2 Mean Field Game on Graph

The MFG theory can be efficiently applied to the problems where there are a large number of players or agents. More specifically, it can reduce the computational complexity of the problem, especially when the number of players is extremely large compared with the traditional game theory framework, such as the classical evolutionary game or Nash non-cooperative game. The single player can update its strategy through interacting with the collective behavior (i.e., mean field) of other players in the game. Since the MEC can be implemented by a large number of users or tasks in which the individual user or task can have an impact on the whole system, we propose the MFG to formulate the task placement problem in the cooperative MEC. The detailed presentation of MFG on the graph is shown below.

Each user is an individual agent in the MFG. Assume that all of the MEC servers make up a graph, and the tasks can be scheduled to the adjacent MEC servers from the current server in one scheduling step. Let $\mathcal{G}(\mathcal{V}, \mathcal{E})$ be a strategy graph. In the strategy graph, the vertex set $\mathcal{V} = \{1, 2, \ldots, D\}$ represents all possible states (MEC servers of the users, which means each task can be allocated to D different servers; the edge set $\mathcal{E}$ consists of all possible transition paths between servers (the task can be scheduled from servers m to n only if $(m, n) \in \mathcal{E}$). For each vertex m, we define the possible transaction situation as $\mathcal{V}_m := \{n : (n, m) \in \mathcal{E}\}$. By considering the current vertex itself, we have $\overline{\mathcal{V}}_m := \mathcal{V}_m \cup \{m\}$. This means the task can be placed to the servers $n \in \overline{\mathcal{V}}_m$ on the next time step if the current server is m. Let s_m be the proportion of users in state m, i.e., the proportion of users who are assigned to run the task in server m. Additionally, $\mathbf{s} := (s_1, s_2, \ldots, s_D)$ is same as the proportion s in (5.5). According to the optimal strategy (transition probability) found by MFG, the task distribution s will change to minimize cost L. In this way, the major components of MFG on the graph can be defined as

- **Population distribution** s^t. s^t_m represents the fraction of users whose tasks are executed on MEC server m at time step t.
- **Transition probability** a^t. a^t_{mn} is the probability that the task in server m is transferred to server n for execution at time t. Further, a^t_m is the *action* of users in server m. a_{mn} equals to 0 if servers m and n are not connected. The forward Fokker-Planck (FPK) equation for each server n can be defined as

$$s_n^{t+1} = \sum_{m \in \overline{\mathcal{V}}_n} a^t_{mn} s^t_m. \tag{5.6}$$

- **Reward** $r_m(s^t, a^t_m) := -\sum_{n \in \overline{\mathcal{V}}_m} a^t_{mn} r_{mn}(s^t, a^t_m)$. This is the instant reward received by the users whose tasks are executed on servers m with population distribution s^t and choose action a^t_m. The reward r_{mn} can be defined as the negative of the average cost for users whose tasks are placed from server m to n at time t according to (5.3), and r_m can be seen as the negative of the expected cost for the users in server m. Moreover, the reward r of all the users in the system is the negative of the cost function $-L$ as shown in (5.5) at time t.
- **Value function** V^t. The value function represents expected cumulative rewards at time t. The value function of each server m is defined as below, which is also the backward Hamilton-Jacobi-Bellman (HJB) equation in MFG

$$\begin{aligned} V^t_m &= \max_{a^t_m} \left\{ r_m\left(s^t, a^t_m\right) + \sum_{n \in \overline{\mathcal{V}}_m} a^t_{mn} V^{t+1}_n \right\} \\ &= \max_{a^t_m} \left\{ \sum_{n \in \overline{\mathcal{V}}_m} a^t_{mn} r_{mn}(s^t, a^t_m) + \sum_{n \in \overline{\mathcal{V}}_m} a^t_{mn} V^{t+1}_n \right\}. \end{aligned} \tag{5.7}$$

According to the major components shown above, the Nash equilibrium can be obtained as the solution of our MFG on graph.

Definition 5.1 In the above description, the Nash equilibrium can be defined as below with the optimal action a^*

$$-L(s, a^*) \geqslant -L(s, \bar{\mathcal{A}}(\bar{a^*})) \tag{5.8}$$

$$L(s, a^*) \leqslant L(s, \bar{\mathcal{A}}(\bar{a^*})). \tag{5.9}$$

$\bar{\mathcal{A}}(\bar{a^*})$ is the action space of all possible actions except optimal action a^*. This means at distribution s, the rewards of all users (cost L) with optimal action a^* cannot be decreased by changing to other actions $\bar{a^*} \in \bar{\mathcal{A}}$. The solution of this MFG is to find the optimal action at each step, which is satisfied with FPK (5.6) and HJB (5.7).

5.3 Mean Field Model to Markov Decision Process

In this section, we prove that the above MFG problem on graph can be converted to the single-agent MDP. State trajectory achieved by a series of actions in the MDP is in accordance with the state dynamics in the MFG, that is, FPK equation in (5.6). This can lead to the conclusion that the MDP and the MFG have the same goal, and solving the above MFG is equivalent to find the optimal solution for the MDP. For this MFG on graph, it can be changed to a RL problem that fits within the Markov properties. It is important to note that in the above MFG problem, the generic users obtain their optimal strategy with the information of population density. Therefore, it can be converted to a single-agent RL problem, which means our proposed task placement system is a central-assisted. We will prove that the above MFG problem can be converted to a single-agent RL model in the following parts of this section. The definition of the RL problem is given by

- **State** $s^t \in \mathcal{S}$. The distribution of all the users at time t

$$s^t = (s_1^t, s_2^t, \ldots, s_D^t), \tag{5.10}$$

  where D is the number of MEC servers.
- **Action** $a^t \in \mathcal{A}$. The transaction probability at time t. The action is the proportion of the users whose tasks in the current server switching to other servers (include the current server itself) at the next moment. It can represent the flow of tasks among the servers

$$a^t = (a_1^t, a_2^t, \ldots, a_N^t). \tag{5.11}$$

  There will be N strategies in each action at time t. We redefine the index of each strategy for a more convenient way to explain the physical meaning of the action. It is related to the shape of the graph and the connection relationship between the servers. Specifically, N is equal to the number of undirected connection paths (edges) in the graph. For instance, if D servers are linearly arranged, which means each server has two adjacent servers. In this case, N equals to $D - 1$, i.e., there are N kinds of connections (edges) in our graph.

- **Reward** $r(s^t, a^t)$. It is the reward received by all the users with the state s^t and action a^t, which is equal to the negative of the cost $-L$ at time t. Our goal is to maximize the reward, that is, minimize the cost L

$$r(s^t, a^t) = -\sum_{m=1}^{D} s_m^t \sum_{n \in \overline{\mathcal{V}}_m} a_{mn}^t r_{mn}(s^t, a_m^t) = -L^t. \tag{5.12}$$

We now define $V^*(s^t)$ as the weighted sum of all the value functions defined in MFG for the servers. Similarly to [199], we show that it is the value function of the above MDP problem, which will be satisfied with the Bellman optimally equation. Once $V^*(s^t)$ can be proved to satisfy the Bellman optimally equation, we can say that the proposed MFG problem can be converted to the above single-agent MDP. The proof can be described as

$$\begin{aligned}
V^*(s^t) &= \sum_{m=1}^{D} s_m^t V_m^t \\
&= \sum_{m=1}^{D} s_m^t \max_{a_m^t} \left\{ \sum_{n \in \overline{\mathcal{V}}_m} a_{mn}^t r_{mn}(s^t, a_m^t) + \sum_{n \in \overline{\mathcal{V}}_m} a_{mn}^t V_n^{t+1} \right\} \\
&= \max_{a^t} \left\{ \sum_{m=1}^{D} s_m^t r_m(s^t, a_m^t) + \sum_{m=1}^{D} \sum_{n=1}^{D} a_{mn}^t s_m^t V_n^{t+1} \right\} \\
&= \max_{a^t} \left\{ r(s^t, a^t) + \sum_{n=1}^{D} s_n^{t+1} V_n^{t+1} \right\} \\
&= \max_{a^t} \left\{ r(s^t, a^t) + V^*(s^{t+1}) \right\}.
\end{aligned} \tag{5.13}$$

Therefore, from the above proof, $V^*(s^t)$ satisfies the Bellman optimally equation for the single-agent MDP problem defined in this section. It should be noticed that the above MFG on graph model cannot be converted to a multi-agent MDP. In multi-agent MDP, the value function for one agent is defined only in terms of itself, but the value function for the agent in an MFG explicitly depends on the value functions of other agents [199]. As a result, the state dynamics with the optimal policy solved by the MDP is equivalent to the state trajectory obtained by the Nash equilibrium in MFG. More specifically, at state s, the actions a under the optimal policy in the MDP are the same as the optimal policy a^* with the Nash equilibrium in the MFG.

5.4 Mean Field Game Guided Deep Reinforcement Learning Approach

In this section, an approach combining the RL and the MFG is proposed. More specifically, the RL approach that doesn't need to know the state transition function is given to solve the MDP. Then, MFG-Q that can be obtained by leveraging the evolutionary dynamics in the MFG is defined to solve the task placement problem in the cooperative MEC.

Reinforcement Learning with Nash Equilibrium

First of all, the general RL problems with the Nash equilibrium in game theory are introduced. Subsequently, MFG-Q is defined based on the Nash equilibrium strategies gained in MFG.

In the RL or the MDP problem, with the policy $\pi := \{a^1, a^2, \ldots a^t\}$, the agent will take action a in state s at each time step t. For one stage, policy π is a probability of choosing an action, which can be represented as $\pi(a|s) \in [0, 1]$. After taking the action a, the agent will move to the next state s' and receive the reward r. The agent's object is to maximize the state value function, which can be written as the sum of the expected discounted rewards

$$V_\pi(s) = \sum_{t=0}^{\infty} \gamma^t \mathbb{E}[r^t|\pi, s_0 = s], \tag{5.14}$$

where $\gamma \in (0, 1]$ is the discount factor, and s_0 is the initial state. It represents the value of state s under the policy π.

For the Q-learning algorithm, the state-action value (Q-value) can be defined as the value function, which follows the Bellman equation. The optimal Q-value which is the sum of the expected discounted reward by following the optimal policy π^*, which can be represented as

$$Q^*(s, a) = r(s, a) + \gamma \mathbb{E}_{s'}[V_{\pi^*}(s')]. \tag{5.15}$$

where $r(s, a)$ is the reward for taking action a at state s, s' is the state of the next time step. The value function can be expressed as the expected of the Q-values in Eq. (5.15), which is

$$V_\pi(s) = \mathbb{E}_a[Q_\pi(s, a)], V_{\pi^*}(s) = \mathbb{E}_a[Q^*(s, a)]. \tag{5.16}$$

The optimal policy π^* can be found by applying value-based frameworks (such as Q-learning, Deep Q-Network) or policy-based frameworks (such as Actor-Critic approach) if we know the optimal Q value $Q^*(s, a)$. The definition of optimal Q-value establishes a relationship between the current time step and the next time step. Moreover, the action is the deterministic policy not the stochastic policy in the Q-learning algorithm. Hence, the optimal state value can be redefined as $V_{\pi^*}(s) =$

$\max_a Q(s, a)$. According to this property, Q-learning initializes its Q-value, and update the Q-value by the following equation

$$\begin{aligned} Q^{t+1}(s,a) =& (1-\alpha)Q^t(s,a) + \alpha[r^t + \gamma \max_{a'} Q^t(s',a')] \\ =& Q^t(s,a) + \alpha[r^t + \gamma \max_{a'} Q^t(s',a') - Q^t(s,a)], \end{aligned} \tag{5.17}$$

where α is the learning rate. $r^t + \gamma \max_{a'} Q^t(s', a')$ is known as the temporal difference (TD) target, where $\max_{a'} Q^t(s', a')$ is called the target Q-value. It takes the difference between the actual and predicted values, and we call it the temporal difference (TD) error. Moreover, it can be proved that the Q-value will converge to the optimal Q-value if all the states and actions can be visited infinitely.

Combining game theory with the RL framework, the optimal policy π^* can be seen as the Nash equilibrium in game theory. Therefore, the Nash equilibrium is the polices π^*, which can be defined as

$$V_{\pi^*}(s) \geqslant V_\pi(s) \text{ for all } \pi \in \Pi, \tag{5.18}$$

where Π is the set of all possible policy in state s.

By considering the Nash equilibrium in game theory. Hu et al. [91] proposed a new optimal Q-value estimation method, where they replace the target Q-value with Nash equilibrium solved by the game theory at each stage. This update estimation is called Nash Q-value, which is defined by the following equation

$$Nash\ Q^*(s,a) = r(s,a) + \gamma \mathbb{E}_{s'}[Nash\ V_{\pi^*}(s')]. \tag{5.19}$$

The optimal policy π^* is the Nash equilibrium which can be gained through the game theory solution. Hu et al. [91] also proves the convergence of Q-value to an equilibrium Nash Q-value.

Similar to the Nash Q-value, we define the mean field guided Q-value (MFG-Q) $Q_{MF}(s, a)$ which can be obtained by solving the MFG problem. We have both the MFG and RL approach in our scheme. In the learning process, the agent takes the action a, gets the reward r and moves to the new state s'. After that, the mean field Nash equilibrium $\pi^*(s')$ is calculated according to the MFG for computing the mean field guided state value ($V_{MF}(s')$) in RL approach. Based on this mean field guided state value ($V_{MF}(s')$), the learning agent will select the action according to the RL approach. Thus, the new update rule which considers the mean field Nash equilibrium for Q-value and TD error can be defined as

$$Q^{t+1}(s,a) = (1-\alpha)Q^t(s,a) + \alpha[r^t + \gamma {V_{MF}}^t(s')], \tag{5.20}$$

where

$$V_{MF}(s') = \mathbb{E}_{a'}[\pi^* Q_{MF}(s',a')] = \sum_{a'} \pi^*(a'|s') Q_{MF}(s',a'). \tag{5.21}$$

In each time step, the agent will update the Q-value according to (5.20) and (5.21). In this way, the key target of the RL problem is how to find the optimal policy π^*, which can be seen as the learning guide of the agent. Since optimal policy π is the mean field equilibrium point in the game theory, we will provide a solution of how to get this equilibrium in the next subsection.

Like the (5.15) and (5.19), we define the mean field guided Q-value (MFG-Q) $Q_{MF}(s, a)$ in the form of

$$Q_{MF}(s, a) = r(s, a) + \gamma \mathbb{E}_{s'}[V_{MF}(s')]. \tag{5.22}$$

It is also an improvement of the estimation for the optimal Q-value. Different from the general RL problem where the estimation of optimal Q-value is gained according to the existing action patterns, our MFG-Q is gained by assuming the equilibrium the agent can reach with the learning environment. By the guidance obtained from the MFG, the direction for updating the Q-value is more clear. Therefore, the convergence rates of the learning process will be significantly improved. More specifically, our MFG-Q is more suitable for the problems that follow the mean field property, i.e., the model can be formulated with a large number of populations, such as the task placement.

5.5 Mean Field Evolutionary Dynamics Approach

Section 5.4 provides the framework for DRL solution. Moreover, it needs the Nash equilibrium obtained by the MFG solution. Thus, the problem is converted to find the optimal policy based on the MFG solution for the RL in each stage. In this subsection, based on the problem formulation in Sect. 5.2, we leverage evolutionary dynamics in the MFG to find the Nash equilibrium.

In Sect. 5.2, each user's task will be assigned to one of the servers to execute. Thus, the population distribution, which is the strategy of users in the MFG can be defined as

$$\mathcal{P}(\mathcal{S}) = \{(s_j)_{j=1}^{D} : \sum_{j=1}^{D} s_j = 1, s_j \geqslant 0, j \in \mathcal{D}\}, \tag{5.23}$$

where s_j is the proportion of users' tasks placed in server j. Based on the knowledge of the population game and the MFG, the mean field evolutionary dynamics [38] is defined by the following Theorem.

Theorem 5.1 *Suppose that a population mean field game on the graph $\mathcal{G}(\mathcal{V}, \mathcal{E})$, the constant $\beta \geqslant 0$ and the payoff function $F : \mathcal{P}(\mathcal{S}) \to \mathbb{R}^n$ are continuous. The evolutionary dynamics is given by*

$$\frac{\mathrm{d}s_j}{\mathrm{d}t} = \sum_{n\in\overline{\mathcal{V}}_j} \frac{1}{d_n} s_n \left[F_j(s) - F_n(s) + \beta\left(\log s_n - \log s_j\right)\right]^+$$
$$- \sum_{n\in\overline{\mathcal{V}}_j} \frac{1}{d_j} s_j \left[F_n(s) - F_j(s) + \beta\left(\log s_j - \log s_n\right)\right]^+, \qquad (5.24)$$

where $\beta \geqslant 0$ is the strength of uncertainty, $[\cdot]^+ = max\{\cdot, 0\}$. Same as the definition in Sect. 5.2, n is the index of the neighboring servers of server j. $d_j = \sum_{n\in\overline{\mathcal{V}}_j} 1$ is the number of the neighboring servers, which represents the degree of graph on vertex j. It can be seen as the forward Fokker-Planck (FPK) equation in MFG, which is also the dynamics of the users' distribution.

Considering the scenario of our task placement in the cooperative MEC problem, we regard the potential as L which is given in (5.5). Accordingly, the payoff function F of the potential L can be defined as

$$F(j) = \frac{\bar{\lambda}\mu_j}{(\mu_j - s_j U\bar{\lambda})^2} + \zeta\kappa_j\bar{\lambda}(\mu_j)^2,\ \forall j \in D. \qquad (5.25)$$

Chow et al. [38] proved that minimizing each player's payoff is equivalent to minimizing the potential. As the MFG model described in Sect. 5.2. Our goal is to achieve a better state of user distribution s by finding an optimal action (transaction probability a), which can minimize the cost L. To obtain the optimal distribution s, a gradient matrix $\Lambda \in \mathbb{R}^{D\times D}$ is constructed to identify the action a, which helps the users make up the new distribution.

We consider the scenario that MEC servers are arranged on the side of the road. In this situation, the MEC servers are linearly arranged, i.e., the tasks in the current server can only be placed to the neighboring two servers in the next time step. Thus, our strategy graph is a one-dimension graph, and each vertex node (server) only has two edges (connection relationship) to the adjacent servers. In each time step, the distribution of users will change based on the task placement in adjacent servers. We define that the gradient matrix $\Lambda \in \mathbb{R}^{D\times D}$ is a population flow matrix, and Λdt represents the proportion of the flows, which can be seen as the action a as described in Sect. 5.2, where dt is the size of the step. For example, $a_{mn} = \Lambda(m,n)dt = p$ means that p percent of the tasks in server m will be placed to server n according to our strategy.

More specifically, through Theorem. 5.1, gradient matrix $\Lambda \in \mathbb{R}^{D\times D}$ will be computed according to the following equations. Due to server 1 and server D are only connected to one neighboring server, the gradients related to them can be given as

$$\Lambda(1,1) = -[F(1) - F(2)]^+,$$
$$\Lambda(1,2) = [F(2) - F(1)]^+,$$
$$\Lambda(D,D) = -[F(D) - F(D-1)]^+,$$
$$\Lambda(D,D-1) = [F(D-1) - F(D)]^+. \tag{5.26}$$

For the other servers $j : \{j \in \mathcal{D} : j \neq 1, D\}$, they have connections to server $j - 1$ and $j + 1$. Therefore, the related gradients can be computed as

$$\Lambda(j,j) = -[F(j) - F(j+1)]^+ - [F(j) - F(j-1)]^+,$$
$$\Lambda(j,j+1) = [F(j+1) - F(j)]^+,$$
$$\Lambda(j,j-1) = [F(j-1) - F(j)]^+. \tag{5.27}$$

During iteration t, the new distribution s^t is updated by distribution s^{t-1} from previous iteration and the gradient matrix Λ

$$s^t = s^{t-1} + \Lambda^t s^{t-1} dt. \tag{5.28}$$

This solution is called the mean field evolutionary dynamics (MFED) solution. Back to our RL problem described in Sect. 5.4, the mean field guided Q-value (MFG-Q) $Q_{MF}(s, a)$ can be obtained from this MFG solution. In Eq. (5.21), s' is the new distribution we get, and a' can be represented as $\Lambda^t dt$.

5.6 Implementation and Overall Architecture

From the perspective of the RL approach, due to the high dimension of action space (the task placement strategies are unlimited) and state space (users' distribution status are unlimited) for our task placement in the cooperative MEC, the DRL framework is considered to solve the dimensional explosion problem. We exploit a new DRL approach with the deep deterministic policy gradient (DDPG) [130] approach, an Actor-Critic scheme as our foundation. The framework of our MFG guided DRL approach is shown in Fig. 5.1. In one state, the servers interact with the MEC environment (transferring the users' tasks among the servers) and get the instant reward. After that, the critic network calculates the TD error according to the instant reward and the optimal action through the MFG solution. The MFG solution can also be obtained from the existing MEC environment. This TD error guides the actor network to update its parameters and generate a new action.

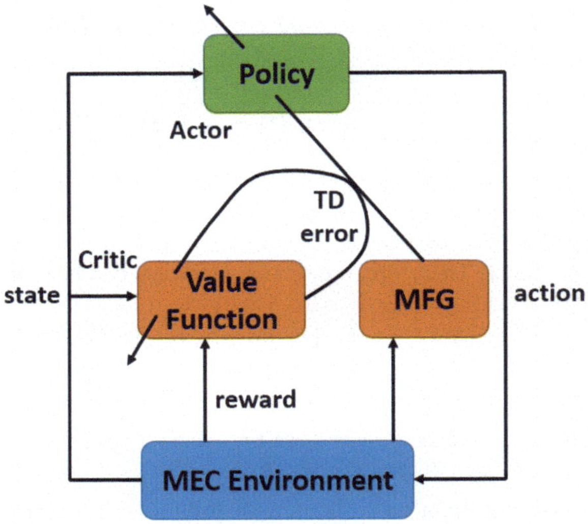

Fig. 5.1 Framework of proposed MFG guided DRL approach

In the **critic network**, we apply the neural network with the parameter ω as the function approximator to represent the Q-value. The updated rule in Eq. (5.20) can be reformulated with the parameter ω. Similar to the Deep Q-Network (DQN) [148], the *Loss* function of the critic network with the MFG-Q can be defined as

$$\mathbb{L} = (y - Q^{\omega}(s, a))^2, \tag{5.29}$$

where $y = r(s, a) + \gamma {V_{MF}}^{\omega'}(s')$ is the target mean field value with the parameter ω'. We can minimize the *Loss* function by applying the gradient based optimization method, i.e., mini-batch stochastic gradient descent.

Different from the value-based framework, we can obtain the policy from the **actor network** $\Omega(s)$ with parameter θ directly. The target of actor network is to find the policy which can maximize the objective function J. The actor network is trained by using sampled policy gradient, and the gradient can be calculated as

$$\nabla_{\theta} J(\Omega) = \nabla_a Q^{\omega}(s, a)|_{a=\Omega(s)} \nabla_{\theta} \Omega^{\theta}(s). \tag{5.30}$$

Different from DDPG, we only have one actor network, which means we don't have the target actor network. We don't need the target action from the target actor network to update the critic network. We directly apply the TD error with Nash equilibrium solved by the MFG as described in (5.20) to update the actor and critic networks.

Algorithm 4 Mean field game guided deep reinforcement learning algorithm

Input: Randomly initialize the critic network $Q^{\omega}(s, a)$ and actor network $\Omega^{\theta}(s)$ with parameter ω and θ separately. Initialize the replay memory $\mathbb{M}$ and set the parameters of target network Q' with the parameter ω' by copying from Q.

Output: the selected action for all of the tasks.

1: **for** episode $=1, \ldots, N$ **do**
2: Initialize the noise process $\mathcal{N}$ for action exploration.
3: The users' distribution at the state s_1.
4: **for** $t = 1, \ldots, T$ **do**
5: Select action $a_t = \Omega^{\theta}(s^t) + \mathcal{N}_t$ according to the actor network and the noise process.
6: The population executes the action a^t, gets the immediate reward r^t, and observes the next state s^{t+1}.
7: Store transition (s^t, a^t, r^t, s^{t+1}) in memory.
8: Randomly sample mini-batch size M of transitions (s_j, a_j, r_j, s_{j+1}) in memory.
9: Set Q value as $y_j = r_j + \gamma V_{MF}{}^{\omega'}(s_{j+1})$ by (5.21), where the optimal policy can be obtained by (5.28).
10: Update the Critic network by minimizing the Loss function $\mathbb{L} = \frac{1}{M}\sum_j (y_j - Q^{\omega}(s_j, a_j))^2$ according to Adam Algorithm.
11: Update the actor network by sampled policy gradient with $\nabla_{\theta} J(\Omega) = \frac{1}{M}\sum_j \nabla_a Q^{\omega}(s, a)|_{a=\Omega(s_j)} \nabla_{\theta}\Omega^{\theta}(s_j)$.
12: Update the parameter ω' in target networks Q' in "soft" way: $\omega' \leftarrow \tau\omega + (1-\tau)\omega'$
13: **end for**
14: **end for**

5.7 Performance Evaluation

The pseudocode for our MFG guided DRL solution is manifested in Algorithm 4. First of all, the users' population interacts with the MEC environment from step 1 to step 6. Moreover, we introduce the noise process $\mathcal{N}$ to balance the exploration and explanation, and $\mathcal{N}$ is the Gaussian noise. In step 9, we set the label of critic network according to (5.21) with the parameter ω'. Further, the optimal policy for calculating label can be gained by solving the (5.28), which is related to the solution of the MFG. In this situation, different from the DDPG method, the process of calculating TD error in our approach contains the contribution of MFG as shown in Fig. 5.1. Parameters updating for critic and actor network are shown in steps 10 and 11. Finally, the parameters of the target network Q' are updated by copying from the critic network Q according to a "soft" way [130]. τ is the soft parameter, and this can make the target value change slowly.

To ascertain the performance of our proposed method, we perform the following comprehensive simulations. We consider the problem that MEC servers are linearly arranged, and each server is connected to two adjacent servers. This problem is very applicable to the scene of the base station arranged along the road. Accordingly, we obtain the real-world user distribution along the road from 4G eNBs dataset of the Internet service provider. The dataset contains network traffic data collected from 3072 4G eNBs in a southern city of China in 2015.

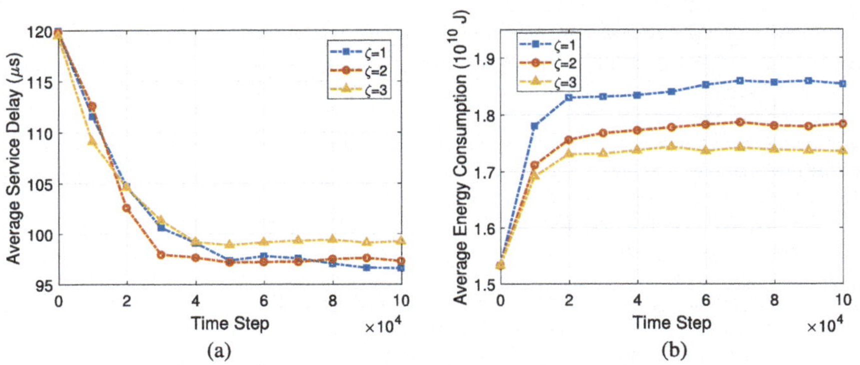

Fig. 5.2 Adjust balance factor ζ. (**a**) Service delay reduction. (**b**) Energy consumption enhancing

We select one month period and obtain the users' distribution at different times from the dataset as our initial users' distributions. At this time, the performance of the task placement with the initial users distribution can be seen as the baseline. All of the users are generating the computations according to the Poisson process with the mean rate $\bar{\lambda} = 3 \times 10^7$ cycle. Assume that there are around 1×10^4 users, and we place 10 MEC servers along the road for service where the service rate for each MEC server belongs to $\mu \in (300, 400)$ GHz. We set the effective capacitance coefficient κ_j belongs to (0.8, 1.2) $(10^{-20}\,\mathrm{J} \cdot \mathrm{s}^2/\mathrm{cycle}^3)$ due to the different CPU architectures of servers. In our implementation, we adopt the fully connected neural networks with two hidden layers for our actor and critic networks, where each hidden layer has 128 nodes. For the neural network training process, we set the experience replay memory size as 2000, the batch size as 32, and the learning rate as 0.001. For the RL part, discount factor γ is 0.9 and "soft" parameter τ is 0.01. Particularly, all of the parameters of the learning approach are got from parameter tuning.

Our goal is to reduce the average service delay. In the process of reducing the delay, the energy consumption will increase. Therefore, we want to ensure that energy consumption does not increase too much according to our approach. The balance factor ζ in Eq. (5.5) is adjusted to achieve this goal. To choose the best value of the balance factor ζ, we evaluate the performance during changing the different values of ζ under a fixed-time user initial distribution. We set ζ as $1, 2, 3 \times 10^{-8}$ to ensure that delay and energy are on the same scale. To avoid the high variance of the learning process, we take five experiments and the average evaluation results are shown in Fig. 5.2. Each step represents one iteration. We find that when ζ equals to 2, the service delay in Fig. 5.2a reduces significantly, and the energy consumption in Fig. 5.2b doesn't increase too much. When ζ equals to 1, the delay reduction 19.16% is very similar to 18.75% when $\zeta = 2$, but the increasing in energy consumption is huge, and this imbalance also happens when $\zeta = 3$. Therefore, we select $\zeta = 2$ for further evaluation.

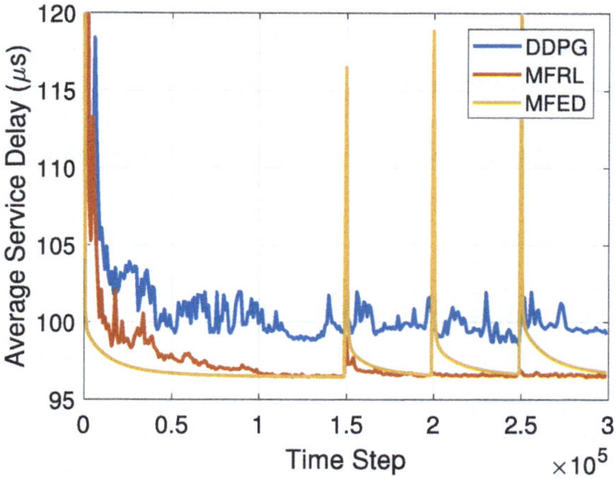

Fig. 5.3 Compare with DRL and MFG method

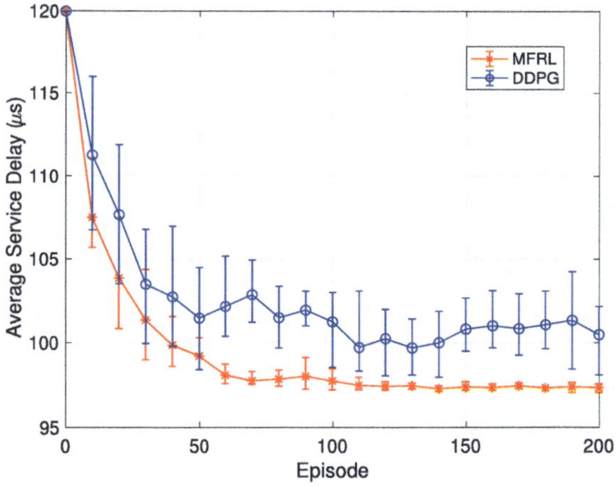

Fig. 5.4 Compare MFRL with DDPG in dynamic environment

We compare our MFG guided DRL approach with other approaches to verify the effectiveness of our proposed scheme. To facilitate display performance, we temporarily ignore the penalty term "energy consumption" and only show the performance of service delay. The evaluation results are shown in Fig. 5.3. We mainly focus on service delay reduction. Comparing to the pure DRL method deep deterministic policy gradient (DDPG) and pure MFG solution [69] Mean field evolutionary dynamics (MFED), our approach (MFRL) has the faster convergence rates and less decision making delay, as shown in Fig. 5.4. We show four stages of the training process in Fig. 5.3, that is, at time step $(0, 1.5, 2, 2.5) \times 10^5$, we have the new user distribution to do the task placement. Different from the DDPG which is not stable and cannot converge to the optimal solution (MFED solution), our MFRL approach converges quickly and approaches the optimal value easily. Even though at the initial stage of the training, our approach still converges quickly, which can speed up the learning process effectively. Besides, the MFED (or other optimization approaches) can only solve the problem when the initial distribution is certain. As shown by the yellow line in Fig. 5.3, MFED needs to resolve the problem from scratch when the new distribution is coming. However, our approach has the memory of the existing communication environment. When the stage comes to the new distribution, the learning agent can make the decision quickly, and won't lead to intolerable decision-making delay. At the same time, our MFRL approach can converge to the point which is very similar to the MFED solution.

5.8 Chapter Summary

To reduce the service delay in the cooperative MEC, we have developed a mean field game guided deep reinforcement learning approach to do the task placement. We have ensured that the MFG and the DRL approach had the common objective. Furthermore, based on the mean filed guided Qvalue (MFG-Q) we proposed obtained by the Nash equilibrium in the MFG, the learning directions were more clear of our approach compared to the pure DRL approach. The multiple simulations have shown that our MFG guided DRL approach was able to reduce service delay by more than 20% and had a significant performance improvement over pure MFG or pure DRL approach. On the one hand, in comparison with pure DRL approach, apart from reducing much service delay, our approach was more stable and had the faster convergence rates during learning; On the other hand, different from the MFG where we need to resolve the problem for each tasks distribution, our approach could make task placement decisions directly when training is enough. Consequently, we could make the task placement decision timely.

Chapter 6
Incentive Mechanism Design in Satellite-Based Federated Learning Using Mean Field Evolutionary Approach

With the increasing development of Low-Earth-Orbit (LEO) Satelite Communications (SatComs), it is foreseen that they will play an important role in broadening the horizon of Federated Learning (FL). Specifically, SatComs can amplify FL by providing consistent global transmission, bridging terrestrial network gaps, and ensuring robust, reliable connectivity in remote or challenging terrains. In this chapter, we consider a SatComs-based FL framework, where satellites in the low-earth orbit collaborate to serve as global servers, able to collect and aggregate FL model parameters transmitted from the mobile devices on the ground continuously. We investigate the joint server selection and handover design optimization problem for SatComs-based FL from the perspective of the system's energy consumption and performance. To address the scalability issue, we propose a Mean-Field-Evolutionary (MFEv) approach that simplifies the interaction between mobile devices as a distribution over their state space, known as the mean-field approximation. It iteratively updates devices' strategies along the Fokker-Planck gradient flow in satellites' strategy space. Our approach's complexity is linear in the number of mobile devices, and we prove that it converges to a unique optimal solution. Numerical simulations demonstrate that our approach is effective in economically benefiting the system and reducing algorithm running time.

6.1 Introduction

The continually increasing data generated at edge devices stems from billions of interconnected Internet-of-Things (IoT) devices as every active IoT client gathers its observed information and forwards it to the edge. Traditional Machine Learning (ML) methods typically aggregate the gathered data on a central data center or a single machine, and such a centralized learning scheme is common among AI-driven companies such as Google and Microsoft [95]. To enhance model

Y. Kang et al., *Mean Field Guided Machine Learning*, Wireless Networks,
https://doi.org/10.1007/978-3-031-91859-9_6

performance using this centralized approach, users might have to sacrifice their privacy by transmitting personal data to these centers. This training approach can be intrusive to privacy, especially when individuals have to share personal or sensitive information to improve the training model's performance. However, the emergence of Federated Learning (FL) happened to overcome these challenges. FL is an innovative approach to centralized ML where a model is trained across multiple IoT devices or servers while keeping the data localized [145]. Instead of sending the data to a central server for training, FL sends the model to the IoT devices. Each device computes an update to the model based on its local data, and then sends only this update back to the central server, where it is aggregated with updates from other devices to improve the global model.

However, in FL, ensuring consistent and uninterrupted communications is essential to its decentralized mode. Terrestrial communications, including fiber-optic networks, cell towers, and local area networks, provide vast bandwidth in cities but suffer from inherent geographical limitations. (E.g., It can only cover about 20% of the global land area [32].) Satellite Communications (SatComs) emerges as an invaluable alternative in this context, granting global coverage that terrestrial systems cannot match. In the coming years, several ongoing Low Earth Orbit (LEO) constellation initiatives, such as SpaceX Starlink, OneWeb, LeoSat, and Amazon Project Kuiper, are set to launch an impressive fleet of over 46,100 satellites into Earth's orbit [169]. By leveraging satellite communication systems, FL can achieve a truly global scale, incorporating data nodes from every corner of the Earth [50, 113, 202]. Additionally, in scenarios where timely data transmission is crucial, such as disaster response or remote surveillance, satellite communication offers reliability that terrestrial alternatives might not guarantee, especially when ground infrastructure is compromised or non-existent. Moreover, a typical LEO satellite constellation is deployed at a height of 500–2000 km from the Earth, achieving speeds of up to 7.8 km/s, which could offer fast communications (e.g., the propagation latency from a terrestrial base station to an LEO satellite is around 5 ms). This characteristic also enables SatComs to provide highly flexible and rapid FL services in contrast to ground-based stations. Therefore, to realize the full potential of FL in a holistic, global context, satellite communication stands out as an indispensable component.

Motivated by the above observations, we consider in this chapter the SatComs-based FL scheme, where the satellites collaborate to serve as the servers for FL that collect and aggregate FL model parameters transmitted by devices on the ground. Such a SatComs-based FL scheme plays a critical role in supporting several applications. For instance, in maritime operations, ships in remote oceanic expanses rely predominantly on satellite communication, FL can enhance navigation, assess piracy threats, and predict equipment malfunctions [214]. Similarly, in environmental monitoring, this scheme can collate and analyze data from sensors scattered across vast terrains, offering real-time insights into global climate shifts, deforestation, and other ecological phenomena [92]. Remote health monitoring is another beneficiary, where isolated communities can employ FL to process health data locally, forwarding essential insights to centralized hubs for expert evaluation [160]. Further, the proposed scheme also offers significant benefits to other applications such as Aircraft Health Monitoring, Space Exploration, Oil & Gas Exploration, and Disaster Response [126].

Despite the above attractive advantages, the implementation of SatComs in the FL framework faces several technical challenges. The *first challenge* is that the limited battery capacity of mobile devices imposes a limitation on the efficiency and reliability of the entire FL process [95, 151]. Specifically, mobile devices often operate on battery power and have energy constraints. In the context of FL, continuously processing and communicating data can rapidly deplete a device's energy reserve, which leads to interrupted training sessions, inconsistent model updates, and potential loss of critical data. The *second challenge* is that one mobile device can be served by several satellites simultaneously, which means that the server selection strategy should be designed well to meet the energy efficiency and performance requirements of the system considered. The *third challenge* is caused by satellites' high mobility. A typical LEO satellite takes only about 90 minutes to orbit the earth in one round. In view of the high speed of satellites, a scientific handover mechanism between satellites needs to be designed well to handle the situation where a satellite moves out of a ground device's reachable area while the FL model parameters are still being transmitted. Toward that end, we consider in this chapter the SatComs-based FL framework and investigate the corresponding joint server selection and handover design optimization problem from the perspective of energy conservation.

To address such issues, many models and algorithms have been proposed in the past. In [131], the authors have considered an incentive mechanism design problem for a hierarchical FL, and an evolutionary approach has been utilized to optimize cluster-selection strategies for local devices. In [220], a replicator dynamics method has been proposed to adapt devices' training strategies that aim to maximize their individual utilities. A Reconfigurable Intelligent Surface (RIS)-aided hybrid network is proposed in [150] that aims to maximize the achievable rate by optimizing the transmit power and the receive scalar. In [101], a Q-learning-based method has been utilized to solve the resource optimization problem for signal recognition in satellite mobile edge computing with FL. In [149], the authors resort to RIS to achieve efficient and effective wireless connectivity for over-the-air FL that jointly designing transceivers, tuning reflection coefficients, and selecting clients. In [207, 208], the authors have proposed a reinforcement learning-based method to optimize devices' training strategies in a server-clients incentive problem. In [106, 165], a Stackelberg game model has been introduced and studied to find the equilibrium solution for devices' CPU resource allocation strategy.

However, as a collaborative machine learning technique, FL incorporates a certain number of distributed local devices' resources to train a shared model. Given the accelerated trend of FL deployment, in the future, the FL operators would have to deal with a massive number (e.g., over thousands) of mobile devices participating in the FL collaboration simultaneously, where above traditional game models and optimization algorithms that study multiple users to compete for the limited resources can quickly become intractable and computationally expensive [165, 208, 220]. Moreover, existing algorithms such as learning-based methods might not guarantee a global optimal offloading solution in practical scenarios (i.e., they can provide only local optimality) [131, 208, 220]. The need to coordinate a large number of local devices to achieve their potential in improving the FL model performance in a cost-effective manner inspires us to bring in and adopt the framework of the mean-field game (MFG) theory.

MFG theory, first introduced concurrently by Lasry and Lions in [117], and Caines and Huang in [23], studies the behavior of a system that involves a large number of agents, modeled by controlled stochastic dynamics. It has been applied in many fields such as economics, social and behavioral sciences, engineering, physics, and biology. For example, in [103, 104, 212], the authors have studied problems arising in mobile vehicle control and cellular communication networks using MFGs. In [121], the authors have studied the propagation of epidemics via mean-field control. Other selected applications include demand response management in the smart grid [218], the evolution of opinions in social networks [11], molecular biology in transport through membranes [219], and data offloading optimization for large-scale edge computing networks [105]. Compared with conventional game theory-based approaches that focus on the interactions between individual agents, MFG restates the model by studying the collective behavior of all agents summarized as one aggregate quantity (termed *mean-field*) and thus has the potential to scale to extremely large populations of agents. By incorporating MFG theory into the SatComs-based FL framework considered, the primary interactions among the ground devices, such as the resource allocation strategy and the local training strategy, together with the interactions between the local devices and the satellites, such as global model aggregation, and payoff strategies, could be simplified significantly for such a large-scale FL system.

In this chapter, we optimize the joint server selection and handover design strategy for a large-scale SatComs-based FL network from both an energy-efficiency perspective and an FL performance perspective. Specifically, in the system considered, in order to attract as many mobile devices as possible to perform the FL task, the satellites assign a certain amount of bonus pool to mobile devices, and each device competes against the others for the limited bonus payoff. Under the influence of other devices' strategies, each device aims to minimize its energy consumption and maximize its own payoff by optimizing its server selection strategy. On the other hand, the satellites also aim to minimize their own energy consumption by optimizing their handover strategy. The server selection problem and handover design problem are coupled through the system cost function. Contributions of this work are thus summarized as follows.

- We propose a SatComs-based FL framework, study the pros and cons compared with traditional terrestrial-communication-based FL, and investigate suitable applications. and study the joint server selection and handover design problem from the perspective of energy consumption and system performance.
- We study the joint server selection and handover design problem from the perspective of energy consumption and system performance of the proposed SatComs-based FL. To tackle the scalability issue, we propose a novel mean-field evolutionary (MFEv) approach that simplifies the interaction among mobile devices as a distribution over their strategy space.

- A comprehensive theoretical analysis is conducted. The computational complexity of the proposed MFEv algorithm is proved to be linear in the number of mobile devices and satellites.
- Numerical simulations prove the superiority of our approach in reducing the overall system's cost by about 31% and saving code running time by about 80% compared with state-of-the-art algorithms.

The rest of the chapter is organized as follows. In Sect. 6.2, we first introduce the system model of the considered scenario and describe the joint optimization problem. Then we propose a novel MFEv approach to solve the optimization problem in Sect. 6.3. In Sect. 6.4, we present numerical results and provide a comprehensive analysis based on them. Finally, conclusions are drawn in Sect. 6.5.

6.2 System Model and Problem Formulation

6.2.1 System Model Description

We consider a SatComs-based FL system that consists of N Low-Earth-Orbit (LEO) satellites and M mobile devices. Mobile devices access satellites from different service areas, each of which can be seen as a circular area on the Earth's surface. Terrestrial backbone networks are interconnected with external networks (e.g., cloud infrastructure) that handle diverse responsibilities, including resource allocation and mission planning. The sets of satellites and mobile devices are denoted by $\mathcal{N} = \{1, 2, \ldots, n, \ldots, N\}$ and $\mathcal{M} = \{1, 2, \ldots, m, \ldots, M\}$, respectively. The satellites orbiting around the Earth collaborate and serve as the global servers, and the mobile devices on the ground serve as distributed local clients in FL. Specifically, satellites that can communicate with each other through inter-satellite links (ISL) have a particular shared machine-learning task to perform, such as traffic crowdsensing [62], location-based recommendation [196], and pollution monitoring [107]. Mobile devices on the ground train the shared FL model on their own local data and transmit the updated model parameters to the satellites. Due to their high mobility, satellites cover different mobile devices during their movement. A conceptual illustration of the system model is given in Fig. 6.1.

In the SatComs-based FL network considered, the FL model training takes place over K communication rounds to minimize the global loss $F^K(\omega)$, where ω is the global model parameter. During each k-th communication round in the set $k \in \mathcal{K} = \{1, 2, \ldots, k, \ldots, K\}$, the training process consists of the following four steps:

1. Global Model Download: Every mobile device downloads the global model $\omega^{(k)}$ parameters from the satellites.
2. Local Model Update: Each mobile device trains the received global $\omega^{(k)}$ locally.
3. Local Model Parameter Transmission: The mobile devices transmit the model parameter update to their selected satellite.

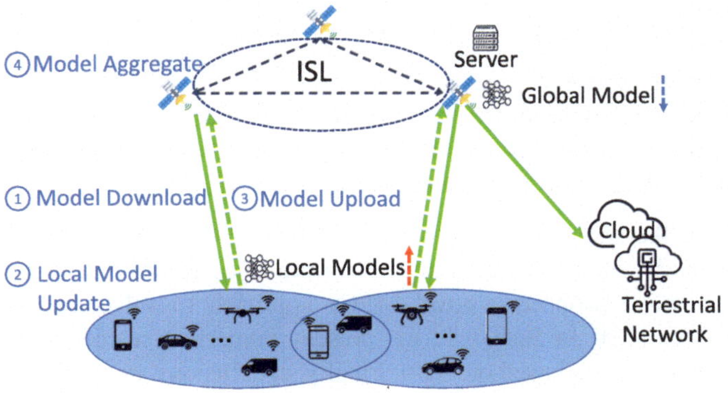

Fig. 6.1 System model of the SatComs-based FL network

4. Global Model Aggregate: After receiving all local model updates received from its mobile devices, each satellite updates the global model of the next communication round as $\omega^{(k+1)}$. An illustrative workflow of the FL process is shown in Fig. 6.1.

Let D_m denote the dataset size of mobile device m and $p_{m,n}^k$ denote the probability that the mobile device m provides local training service to satellite n in the k-th communication round. Let $P_m^k = \{p_{m,1}^k, p_{m,2}^k, \ldots, p_{m,N}^k\}$ denote the satellite selection strategy profile of the mobile device m in the k-th communication round, and $\mathcal{P}^k = \{P_1^k, P_2^k, \ldots, P_m^k, \ldots, P_M^k\}$ denote the set of strategy profiles of all mobile devices. Note that P_m^k is a probability strategy profile, and thus we have:

$$\begin{aligned} & p_{m,n}^k \geq 0, \forall m, n, k \in \mathcal{M}, \mathcal{N}, \mathcal{K}, \\ & \sum_{n=1}^{N} p_{m,n}^k = 1, \forall m, k \in \mathcal{M}, \mathcal{K}. \end{aligned} \tag{6.1}$$

Remark 6.1 It is essential to highlight that our primary here is on optimizing server selection and handover processes within the context of the considered scenario, which falls under the purview of incentive mechanism design. Consequently, our proposed methodology does not influence the convergence of either the FL global model or its local counterparts. As such, any existing convergence analysis such as in [36, 56, 83] remains applicable in our proposed framework.

6.2.2 *Mobile Devices' Payoffs*

In our SatComs-based FL framework considered, satellites assign a certain amount of bonus pool to the mobile devices to attract them, and the devices distribute it among themselves based on their individual contributions. Recall that P_m is mobile device m's satellite selection strategy. Then, the total amount of data offered to satellite n in the k-th communication round is $\sum_{m=1}^{M} D_m p_{m,n}^k$. We quantify the bonus pool as a function of the FL global model performance. According to [208], the test accuracy of the final global model in an FL can be regarded as a log-like concave function $f(.)$ with respect to the amount of all local training data. That is, the shared bonus pool l^k contributed by the satellites in the k-the communication round is quantified as:

$$l^k = f\left(\sum_{m=1}^{M} D_m p_{m,n}^k\right). \tag{6.2}$$

On the other hand, in real-world FL applications, the local data from different mobile devices are usually not independent and not identically distributed (non-iid). To evaluate the contributions of devices fairly, we introduce the contribution index α_m to indicate the data quality of mobile devices [206] which is measured empirically. Therefore, based on Eq. (6.2), the bonus pool shared by satellites in the k-th communication round, denoted as $J_{m,n,k}^{pf}$, is quantified by:

$$J_{m,n,k}^{pf} = f(\sum_{m=1}^{M} \alpha_m D_m p_{m,n}^k) + b_{m,n}^k, \tag{6.3}$$

where $b_{m,n}^k$ is the compensation bonus that satellite n has for device m in the k-th communication round.

When mobile devices receive the bonus pool from the satellites, we assume that their final profits are proportional to their contribution index α_m and the data size that they invest in the FL. That is, the contribution proportion of mobile device m to the satellite n in the k-th communication round is given by: $\frac{\alpha_m D_m p_{m,n}^k}{\sum_{m=1}^{M} \alpha_m D_m p_{m,n}^k}$. Therefore, the final payoff that mobile device m receives from the satellite n in the k-th communication round, denoted by $C_{m,n,k}$, is given as:

$$\begin{aligned} C_{m,n,k}^{pf} =& \lambda_{m,n,k}^{pf} \left(f(\sum_{m=1}^{M} \alpha_m D_m p_{m,n}^k) \right) \\ & \cdot \frac{\alpha_m D_m p_{m,n}^k}{\sum_{m=1}^{M} \alpha_m D_m p_{m,n}^k} + b_{m,n}^k, \end{aligned} \tag{6.4}$$

where $\lambda^{pf}_{m,n,k}$ is the payoff coefficient. It is worthwhile to mention that the local mobile device trains the FL model shared by the server using a mini-batch of its data. Here, D_m refers to the full batch size. The reason that we utilize full-batch D_m in the manuscript is that the mobile devices' payoff is quantified by the devices' full-batch data size, but such quantification will not influence the local training step. In other words, in the proposed scheme, we still use the mini-batch in the local training step in each round of FL.

6.2.3 Mobile Devices' Communication Costs

Energy efficiency has always been a vital metric for large-scale FL frameworks [201]. We consider the wireless transmission energy consumption as one of mobile device m's costs. Based on [209], for a wireless fading channel with a channel gain of g, the transmission energy consumed to transfer s bits of data within a time slot is quantified by a convex monomial function:

$$c = \gamma \frac{s^{\beta}}{g}, \tag{6.5}$$

where γ denotes the energy coefficient, and β denotes the monomial order. γ and β can be obtained from the analytical models or empirical evaluation (e.g., the channel estimation in the real-world wireless channel where the mobile devices and the satellites operate) [9]. Thus, the communication cost of mobile device m for transmitting its updated local model parameters to satellite n in the k-th communication round is given by:

$$C^{cm}_{m,n,k} = \lambda^{cm}_{m,n,k} p^{k}_{m,n} \frac{S^{\beta^{k}_{m,n}}}{g^{k}_{m,n}}, \tag{6.6}$$

where $\lambda^{cm}_{m,n,k}$ is the communication cost coefficient, S is the FL model size shared by the satellites, and $g^{k}_{m,n}$ is the wireless channel gain between the mobile device m and satellite n in the k-th communication round.

6.2.4 Devices' Handover Energy Costs

As shown in Fig. 6.1, different satellites have different coverage areas. When a satellite moves out of a ground device's reachable area, while the device's FL model parameters are still in transmission, then the ground device will need to choose another available satellite to transmit the rest of the model parameters, and such parameters are then transmitted to the first satellite for further integration through the ISL [28], named the *handover mechanism*. However, the onboard battery capacity on satellites is extremely limited since its only energy resource is the solar energy collected by its solar panels [75]. Therefore, we also consider the handover energy consumption of satellite n in the k-th communication round as one of the mobile devices' costs (i.e., mobile devices pay for the handover services provided by satellites). A device's handover cost $C^{hd}_{m,n,k}$ is quantified by:

$$C^{hd}_{m,n,k} = \lambda^{hd}_{m,n,k}\delta^{k}_{m,n}p^{k}_{m,n}E_n H^{k}_{n,n'}\frac{\bar{S}^{k}_{m,n}}{c^{k}_{n,n'}}, \tag{6.7}$$

where $\lambda^{hd}_{m,n,k}$ is the handover energy consumption coefficient. $\delta_{m,n,k}$ is a binary variable capturing whether or not satellite n needs to provide handover service for device m in the k-th communication round:

$$\delta^{k}_{m,n} = \begin{cases} 1, \text{ if satellite } n \text{ needs handover for device } m \\ 0, \text{ other wise.} \end{cases} \tag{6.8}$$

In Eq. (6.7), E_n is a constant transmission power of satellite n in the ISL; $\bar{S}^{k}_{m,n}$ is the size of untransmitted model parameters to be transmitted in the ISL by satellite n from device m; $c^{k}_{n,n'}$ is the transmission capacity in the ISL between satellite n and n'. Further, $H_{n,n'}$ is the satellite's handover strategy that indicates whether or not satellite n transmits the untransmitted model parameters to satellite n' in the k-th communication round:

$$H^{k}_{n,n'} = \begin{cases} 1, \text{ if satellite } n \text{ handover to satellite } n' \\ 0, \text{ other wise.} \end{cases} \tag{6.9}$$

Since satellite n can only choose 1 satellite to transmit the model parameters, we have:

$$\sum_{n' \in \mathcal{V}^{k}_{n}} H^{k}_{n,n'} = 1, \tag{6.10}$$

where $\mathcal{V}^{k}_{n}$ is the set of satellites that builds ISL with satellite n in k-th communication round.

6.2.5 Dynamics of Satellites' Communication Links

In SatComs, a large number of satellites orbit the Earth at very high speeds. As introduced in Sect. 6.2.4, the available satellites set for ground devices changes over time. To capture the dynamics of the available satellite set, we introduce a bipartite

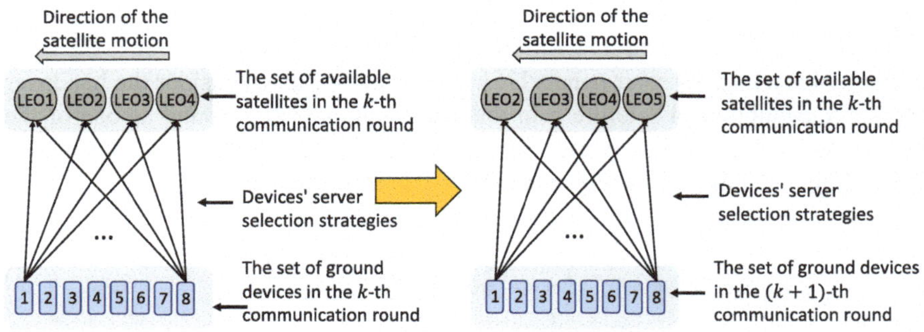

Fig. 6.2 Bipartie graph induced by the dynamics of the available satellites

graph $\langle S_m, \mathcal{B}_n^k, E_{m,n}\rangle$, shown in Fig. 6.2, where the node set $\mathcal{B}_n^k$ at the top side includes the satellites available for the ground devices in the k-th communication round, and the node set at the bottom side includes the mobile devices S_m that attend the FL in the k-th communication round. The edge set $E_{m,n}$ between the satellites set and the devices set represents the devices' server selection strategies. According to [60], in SatComs, satellites orbit the Earth in fully-predictable trajectories. Thus, the available satellite set for every communication round is also fully-predictable. Let $\mathcal{B}_n(k)$ denote the set of available satellites in the k-th communication round. Then:

$$\mathcal{B}_n(k) = \mathcal{B}_n^k, \tag{6.11}$$

where $\mathcal{B}_n^k$ is the predicted available satellite set in the k-th communication round.

6.2.6 Overall Optimization Problems

By summing up the costs and payoffs over total K communication rounds into the SatComs overall system's cost $C_{m,n}$, the overall server selection and handover design optimization problem is given below:

$$\begin{aligned}
&\mathcal{P}_1 : \min_{\mathcal{P},\mathcal{H}} C_{m,n} \\
&= \sum_{k=1}^{K} -C_{m,n,k}^{pf} + C_{m,n,k}^{cm} + C_{m,n,k}^{hd} \\
&= \sum_{k=1}^{K} -\lambda_{m,n,k}^{pf} \left(f(\sum_{m=1}^{M} \alpha_m D_m p_{m,n}^k) + b_{m,n}^k \right) \cdot \frac{\alpha_m D_m p_{m,n}^k}{\sum\limits_{m=1}^{M} \alpha_m D_m p_{m,n}^k}
\end{aligned}$$

$$+ \lambda^{cm}_{m,n} p^k_{m,n} \frac{S^{\beta^k_{m,n}}}{g^k_{m,n}}$$

$$+ \lambda^{hd}_{m,n,k} \delta^k_{m,n} p^k_{m,n} E_n H^k_{n,n'} \frac{\bar{S}^k_{m,n}}{c^k_{n,n'}} \tag{6.12}$$

$$s.t. \quad \begin{cases} C_1 : p^k_{m,n} \geq 0, \\ C_2 : \sum_{n=1}^{N} p^k_{m,n} = 1, \forall m \in \mathcal{M}, \ \forall k \in \{1, 2, \ldots, K\}, \\ C_3 : \mathcal{B}_n(k) = \mathcal{B}^k_n, \end{cases}$$

where $C^{pf}_{m,n}$, $C^{cm}_{m,n}$, $C^{hd}_{m,n}$ are defined in Eqs. (6.4), (6.6), and (6.7), respectively. The constraints C_1, and C_2 come from Eq. (6.1), and C_3 comes from Eq. (6.11). In problem $\mathcal{P}_1$, the devices' and satellites' strategies are coupled through their cost functions. A natural solution concept in this case is the Nash Equilibrium (NE) [10, 90], where no single device has the incentive to modify its strategy unilaterally.

6.3 Mean-Field Evolutionary Approach

In this section, we develop a methodology toward solving the non-cooperative game formulated through Eq. (6.12). We first decompose the game into a discrete MFG game and an Integer programming (IP) problem in Sect. 6.3.1. We show that the IP problem has an analytical optimal solution, and for the discrete MFG problem, we propose a novel mean-field evolutionary approach to find its NE.

6.3.1 Problem Decomposition

Proposition 6.1 *The optimal solution of satellites' handover strategies $H^{k*}_{n,n'}$ can be directly calculated by minimizing the satellites' handover energy cost, which is given as:*

$$H^{k*}_{n,n'} = \begin{cases} 1, & \text{if } c^k_{n,n'} \geq c^k_{n,n''}, \ \forall n'' \in \mathcal{N}, \\ 0, & \text{otherwise}. \end{cases} \tag{6.13}$$

Proof As can be seen in the original game $\mathcal{P}_1$ in Eq. (6.12), satellites' handover strategies $H^k_{n,n'}$ can only influence the handover cost $C^{hd}_{m,n,k}$, rather than the devices' communication costs and payoffs. Therefore, for any given device's server selection strategy $\mathcal{S}$, the optimal value of satellites' handover strategies $H^{k*}_{n,n'}$ can be directly calculated by minimizing the satellites' handover costs (i.e, an Integer Programming (IP) problem):

$$\begin{aligned}\mathcal{P}_2 : \min_{\mathcal{H}} \sum_{k=1}^{K} C^{hd}_{m,n,k} \\ = \sum_{k=1}^{K} \lambda^{hd}_{m,n,k} \delta^{k}_{m,n} p^{k}_{m,n} E_n H^{k}_{n,n'} \frac{\bar{S}^{k}_{m,n}}{c^{k}_{n,n'}}\end{aligned} \tag{6.14}$$

$$s.t. \quad \begin{cases} C_1 : p^{k}_{m,n} \geq 0, \\ C_2 : \sum_{n=1}^{N} p^{k}_{m,n} = 1, \forall m \in \mathcal{M}, \ \forall k \in \{1, 2, \ldots, K\}, \\ C_3 : \mathcal{B}_n(k) = \mathcal{B}^{k}_{n}. \end{cases}$$

This leads to the optimal solution $H^{k*}_{n,n'}$ given in Eq. (6.13). □

6.3.2 Preliminaries of Discrete MFG

When given satellites' optimal handover strategies as in Eq. (6.13), problem P_1 becomes a discrete MFG problem where massive agents (i.e., the mobile devices) interact with others to find their optimal discrete server selection strategies. When the number of agents is large, traditional optimization methods will be intractable to model the interaction between all pairs of agents. In this chapter, we propose a novel MFEv approach to solve the problem $\mathcal{P}_1$ in Eq. (6.12). Before providing an analysis of each step of the MFEv in Sect. 6.3.3, we first introduce the basic idea of discrete MFG.

A discrete MFG is a game played by a large number of players in the player set $\mathcal{M} = \{1, 2, \ldots, M\}$. Each player in the population selects a strategy from the discrete strategy set $\mathcal{N} = \{1, 2, \ldots, N\}$, where N is the size of each player's strategy space. Every player's payoff is dependent on other players' strategies. Thus player m choosing strategy n has the payoff function $C_{m,n} : \mathcal{P}(\mathcal{N}) \to \mathbb{R}$, where $\mathcal{P}(\mathcal{N})$ is defined below.

The aggregated strategy of the population can be described by $P = (p_{m,n})_{n=1}^{N} \in \mathcal{P}(\mathcal{N})$, where $p_{m,n}$ represents the probability of player m choose strategy n, and $\mathcal{P}(\mathcal{N})$ is a probability space (strategy space):

$$\mathcal{P}(\mathcal{N}) = \left\{ (p_{m,n})_{n=1}^{N} \in \mathbb{R}^n : \sum_{n=1}^{N} p_{m,n} = 1, m \in \mathcal{M}, n \in \mathcal{N} \right\}. \tag{6.15}$$

Let $\mathcal{P}_n$ be the set of all possible strategies for player n, and $P = (P_n, P_{-n})$ be a strategy profile, a set consisting of player n's strategy P_n and all the other players' strategies P_{-n}. Based on [123], the game reaches a NE $P^* = (P_n^*, P_{-n}^*)$ if:

$$C_{m,n}(P_n^*, P_{-n}^*) \geq C_{m,n}(P_n, P_{-n}^*), \text{ for all } P_n \in \mathcal{P}_n. \tag{6.16}$$

Given the optimal handover strategy $H_{n,n'}^{k*}$, the optimization problem $\mathcal{P}_1$ in Eq. (6.12) becomes a discrete MFG, where the players are the mobile devices on the ground $\mathcal{M} = \{1, 2, \ldots, M\}$, and their strategies are their server selection strategies $\mathcal{P}^k = \left\{p_{m,n}^k | m \in \mathcal{M}, n \in \mathcal{N}\right\}$ in the k-th communication round. With other players' strategies fixed, every player aims to maximize its own payoff function (minimize cost function in Eq. (6.12)).

Remark 6.2 An MFG game is called a potential game, if there exists a differentiable potential function $F : \mathcal{P}(\mathcal{N}) \to \mathbb{R}$, such that $\frac{\partial F}{\partial P_{m,n}} = C_{m,n}(P)$, for all $n \in \mathcal{N}$. The problem of maximizing each player's payoff is equivalent to maximizing the potential function. Therefore, the discrete MFG problem in Eq. (6.12) is also a potential game.

6.3.3 Mean-Field Evolutionary Approach

In this subsection, we describe the detailed steps of the MFEv approach for the population game in Eq. (6.12). We first construct a strategy graph of the problem $\mathcal{P}_1$, and then define the embedded Riemannian manifold. An MFEv dynamics on the Riemannian manifold is then proposed, and finally, the iterative update strategy is given based on the MFEv dynamics.

6.3.3.1 Strategy Graph Construction with MFG

In Fig. 6.3, two general types of a strategy graph $G = \langle \mathcal{N}, \mathcal{E} \rangle$ of discrete MFG games are shown, where the nodes represent players' strategy set $\mathcal{N}$ (i.e., the set of available satellites in the k-th communication round $\mathcal{B}_n(k)$ in the scenario considered), and an edge in the edge set $\mathcal{E}$ exists between nodes $n, n' \in \mathcal{N}$ if players who currently choose strategy n are able to choose strategy n'. Node n's neighborhood is denoted by:

$$T(n) = \{n' \in \mathcal{N} : (n, n') \in \mathcal{E}\}. \tag{6.17}$$

For many games, players can change from any strategy to another, making the strategy graph G a complete graph, as shown in Fig. 6.3b. However, the strategy graph utilized in the Satellite-based FL considered is the 1-d strategy graph shown in Fig. 6.3a. Although the fully connected strategy graph can also be utilized, the computational complexity of the 1-d strategy graph is much lower.

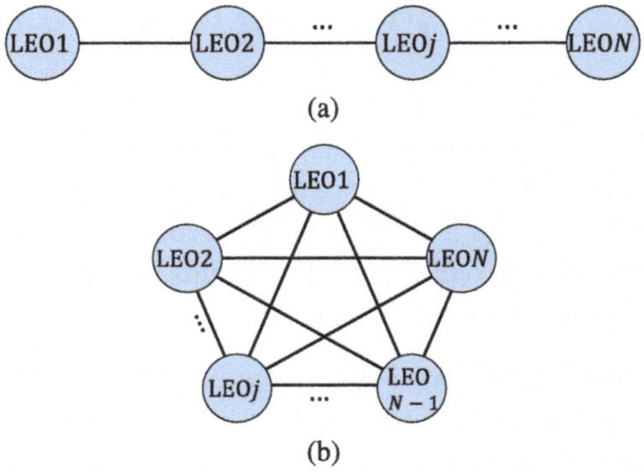

Fig. 6.3 Strategy graph for two types of discrete MFGs. (**a**) 1-dimensional (1-d) strategy graph. (**b**) Fully connected strategy graph

6.3.3.2 Riemannian Manifold Construction

We then construct the Riemannian Manifold $(\mathcal{P}_0(\mathcal{N}), g^W)$ on the strategy graph. Let $\mathcal{P}_0(\mathcal{N})$ denote the interior of $\mathcal{P}(\mathcal{N})$. A Wasserstein metric that induces a Riemannian metric tensor in the interior of probability simplex is employed to measure the distance in the strategy space.

Definition 6.1 Given two discrete probability functions $P^0, P^1 \in \mathcal{P}_0(\mathcal{N})$, the Wasserstin metric W is defined by:

$$W(P^0, P^1)^2 = \inf\left\{ \int_0^1 (\nabla\Phi(t), \nabla\Phi(t))_{P(t)} dt \right.$$

$$\left. : \frac{dP}{dt} + div\,(P\nabla\Phi) = 0, P(0) = P^0, P(1) = P^1 \right\}, \tag{6.18}$$

where $\Phi : \mathcal{N} \to \mathbb{R}$ is a function, and $\nabla\Phi : \mathcal{N} \times \mathcal{N} \to \mathbb{R}$ is:

$$\nabla\Phi_{n,n'} = \begin{cases} \Phi_n - \Phi_{n'}, & \text{if } n, n' \in \mathcal{E}, \\ 0, & \text{otherwise.} \end{cases} \tag{6.19}$$

div is the divergence operator: $div(\mu)_n = -\sum_{n' \in T(n)} \mu_{n,n'}$.

For the purpose of defining our distance function, we use a particular flux function: $\mu_{n,n'} = \theta_{n,n'}(P)\nabla\Phi_{n,n'}$, where $\theta_{n,n'}(P)$ represents the discrete probability of edge (n, n'), defined by:

$$\theta_{n,n'}(P) = \begin{cases} \frac{1}{d_{n'}} P_{n'}, \; F_{m,n'}(P) < F_{m,n}(P), \\ \frac{1}{d_n} P_n, \; F_{m,n'}(\rho) > F_{m,n}(P), \\ \frac{1}{2}\left(\frac{P_n}{d_n} + \frac{P_{n'}}{d_{n'}}\right), \; F_{m,n'}(\rho) = F_{m,n}(P), \end{cases} \tag{6.20}$$

where $F_{m,n} : \mathcal{P}(\mathbb{S}) \rightarrow \mathbb{R}$ is the potential function, and $d_n = \sum_{n' \in T(n)} 1$ is the degree of node n (i.e. the total number of nodes in node n's neighbourhood).

With the Wasserstein metric W, the inner product g_P^W in the Riemannian Manifold is then defined below.

Definition 6.2 For any two tangent vectors $\sigma_1, \sigma_2 \in T_p\mathcal{P}(\mathcal{N})$, the inner product $g_P^W : T_P\mathcal{P}_0(\mathcal{N}) \times T_p\mathcal{P}_0(\mathcal{N}) \rightarrow \mathbb{R}$ is defined by:

$$g_P^W(\sigma^1, \sigma^2) = \frac{1}{2} \sum_{(n,n') \in \mathcal{E}} \theta_{n,n'}(P)(\Phi_n^1 - \Phi_{n'}^1)(\Phi_n^2 - \Phi_{n'}^2), \tag{6.21}$$

where $\sigma^i = -div(\rho \nabla \Phi^i)$ for $i = 1, 2$. $T_P\mathcal{P}_o(\mathcal{N}) = \{(\sigma_n)_{n=1}^N \in \mathbb{R}^N : \sum_{n=1}^N \sigma_n = 0\}$ is the tangent space at a point $P \in \mathcal{P}_o(\mathcal{N})$.

With the strategy space in (6.15) and the inner product in Definition 6.2, we finally construct the Riemannian manifold $(\mathcal{P}_o(\mathcal{N}), g_P^W)$ [38] on which the MFEv dynamics can be defined.

6.3.3.3 Fokker-Planck Equation as MFEv Dynamics

We next define the MFEv dynamics as the Fokker-Planck Equation on the Riemannian manifold.

Theorem 6.1 *Consider a discrete MFG game with strategy graph $G = \langle \mathcal{N}, \mathcal{E} \rangle$, strength of uncertainty constant $\tau > 0$, and a potential function $F : \mathcal{P}(\mathcal{N} \rightarrow \mathbb{R})$. Then, the gradient flow on the constructed Riemannian manifold $\left(\mathcal{P}_0(\mathcal{N}, g^W)\right)$ satisfies the following Fokker-Planck equation:*

$$\frac{\mathrm{d}P_n}{\mathrm{d}t} = \sum_{n' \in T(n)} \frac{1}{d_{n'}} P_{n'}[F_{m,n}(P) - F_{m,n'}(P) + \tau(\log P_{n'} - \log P_n)]^+$$

$$- \sum_{n' \in T(n)} \frac{1}{d_n} P_n[F_{m,n'}(P) - F_{m,n}(P) + \tau(\log P_n - \log PD_{n'})]^+, \tag{6.22}$$

where d_n is the degree of node n, $[\cdot]^+ = \max\{\cdot, 0\}$.

Proof Given the Riemannian manifold $\left(\mathcal{P}_0(\mathcal{N}, g^W)\right)$, for any $\sigma \in T_P\mathcal{P}_0(\mathcal{N})$, there exists Φ such that $\sigma = -div(P\nabla\Phi)$. Since $\frac{dP}{dt} = \left(\frac{dP_n}{dt}\right)_{n=1}^N$ is also in tangent space $T_P\mathcal{P}_o(\mathbb{S})$, by Definition 6.2, we have

$$g_P^W\left(\frac{dP}{dt}, \sigma\right) = \sum_{n=1}^N \frac{dP_n}{dt} \Phi_n. \tag{6.23}$$

On the other hand, define the noisy potential $\bar{F}(P)$ as:

$$\bar{F}(P) = F(P) - \tau \sum_{n=1}^{N} P_n \log P_n, \ \tau > 0, \tag{6.24}$$

where the term $P_n log P_n$ is the Shannon-Boltzmann entropy that models the players' irrational behaviors, making mistakes, or risk-taking behaviors, and τ represents the strength of uncertainty. Based on the definition of the noisy potential $\bar{F}$, we have

$$\begin{aligned} d\bar{F}(P) \cdot \sigma &= \sum_{n=1}^{N} \frac{\partial_n}{\partial P_n} \bar{F}(P) \cdot \sigma_n = -\sum_{n=1}^{N} \bar{F}_n(P) div(P\nabla\Phi)_n \\ &= \left(\nabla\bar{F}(P), \nabla\Phi\right)_P = -\sum_{n=1}^{N} \Phi_n div\left(P\nabla\bar{F}(P)\right)_n. \end{aligned} \tag{6.25}$$

With the definition of gradient flow $d\bar{F}(P)$ on the manifold, and combining Eqs. (6.23) and (6.25), we obtain

$$\begin{aligned} 0 &= g_P^W\left(\frac{dP}{dt}, \sigma\right) - d\bar{F}(P) \cdot \sigma \\ &= \sum_{n=1}^{N} \left\{\frac{dP_n}{dt} + div\left(P\nabla\bar{F}(P)\right)_n\right\} \Phi_n. \end{aligned} \tag{6.26}$$

Noting that the above equation holds for all $(\Phi_n)_{n=1}^{N} \in \mathbb{R}^N$, we finally arrive at

$$\frac{dP_n}{dt} + \sum_{n' \in T(n)} \theta_{n,n'} \left(\bar{F}_{n'}(P) - \bar{F}_n(P)\right) = 0, \tag{6.27}$$

Substituting $\theta_{n,n'}$ with Eq. (6.20) leads to the MFEv dynamics proposed in Eq. (6.22). □

The MFEv dynamics given in Theorem 6.1 constitutes a gradient flow on the Riemannian manifold. Starting from any initial strategy P_0, a final optimal strategy P^* can be reached to optimize players' potential function $F(P)$ with the gradient flow iteration.

6.3.4 Updating Strategies

We will next update the mobile devices' strategy in an iterative manner with the proposed MFEv dynamics. Specifically, let $\Lambda_m \in \mathbb{R}^{N\times N}$ be the gradient matrix of device m. First, we give $F(n)$ as:

$$F_{m,n}(P) = \frac{\partial C_{m,n}}{\partial p_{m,n}^k}, \tag{6.28}$$

As shown in the 1-d dimensional strategies graph in Fig. 6.3a, node 1 and node N only have connection with node 2 and node $(N-1)$, respectively. The gradients are (for $\tau = 0$):

$$\begin{aligned}
\Lambda_m(1,1) &= -[F_{m,1} - F_{m,2}]^+,\\
\Lambda_m(1,2) &= [F_{m,2} - F_{m,1}]^+,\\
\Lambda_m(N,N) &= -[F_{m,N} - F_{m,N-1}]^+,\\
\Lambda_m(N,N-1) &= [F_{m,N-1} - F_{m,N}]^+.
\end{aligned} \tag{6.29}$$

For all the other nodes in the strategy graph, i.e., $\{n | n \in \mathcal{N}, n \neq 1, N\}$, the gradients are (for $\tau = 0$):

$$\begin{aligned}
\Lambda(n,n) &= -[F(n) - F(n+1)]^+ - [F(n) - F(n-1)]^+,\\
\Lambda(n,n+1) &= [F(n+1) - F(n)]^+,\\
\Lambda(n,n-1) &= [F(n-1) - F(n)]^+.
\end{aligned} \tag{6.30}$$

For all the other elements in the gradient matrix Λ_m that are not calculated in Eq. (6.29) and (6.30), their values are set to 0.

Finally, we update mobile devices' strategies in an interactive manner with their gradient matrix Λ_m. Recall $P_m^k = \{p_{m,1}^k, p_{m,2}^k, \ldots, p_{m,N}^k\}$ denotes the satellite selection strategy profile of the mobile device m in the k-th communication round. In the l-th iteration, the gradient matrix $\Lambda_m(l)$ for all $m \in \mathcal{M}$ are first calculated. Then, device m's strategy is updated by

$$P_m^k(l) = P_m^k(l-1) + \Lambda_m(l) P_m^k(l-1) dt, \tag{6.31}$$

where dt is a small time step. The iteration process stops when the following condition is satisfied.

$$\sum_{m=1}^{M} \left| P_m^k(l) - P_m^k(l-1) \right| \leq \delta_0, \tag{6.32}$$

where δ_0 is a small threshold. A summary of the proposed MFEv approach is given in Algorithm 5.

Algorithm 5 Mean-field evolutionary approach for population game in Eq. (6.12)

```
Random initialize all devices' strategy profile 𝒫
while (6.32) is not satisfied do
    for each device m do
        gradient matrix Λ_i construction by (6.29) and (6.30)
        update device m's strategy P_m^k by (6.31)
    end for
end while
Return: strategy profile 𝒫
```

6.3.5 MFEv Performance Analysis

We now finally provide here a comprehensive performance analysis of the proposed MFEv approach.

6.3.5.1 Convergence Analysis

Proposition 6.2 *The solution to the MFG problem in Eq. (6.12) obtained by the proposed MFEv approach is unique.*

Proof Consider a discrete MFG potential game $G = \langle S, E \rangle$, with potential function $F(P)$ and a constant τ. According to [123], for any initial strategy $\mathcal{P}_0$, updating players' strategies following the gradient flow (i.e., Fokker Planck) defined in Eq. (6.22) will converge to a unique solution $\mathcal{P}^*$. Since the discrete MFG game in Eq. (6.12) is potential, we conclude that the solution obtained by the proposed MFEv approach is optimal. □

6.3.5.2 Complexity Analysis

Proposition 6.3 *The time complexity of the proposed MFEv method is $O(MN)$, where M is the number of mobile devices and N is the number of available satellites in every communication round.*

Proof As is shown in Algorithm 5, in order to update devices' strategies, during each iteration, we construct a gradient matrix $\Lambda_{m,n}$ for every mobile device and every satellite as in Eq. (6.22). Therefore, the time complexity for our MFEv approach is $O(MN)$. □

6.4 Simulation Results

In this section, we evaluate the performance of the proposed MFEv scheme from four perspectives: (i) Convergence Analysis; (ii) Performance analysis; (iii) Sensitivity analysis; (iv) Scalability analysis.

6.4.1 *Simulation Scenario Description and Parameters Initialization*

In the simulations, based on the LEO satellites orbit data and SatComs constellation data described in [24], we have utilized the Systems Tool Kit (STK) to simulate the Starlink Constellations. Specifically, the constellation mainly consists of two shells. The first shell includes 1440 satellites at a 550 km altitude, which are made up of 72 orbital planes of 20 satellites each. Each satellite has an inclination of 43°, and an elevation angle of 40°. The second shell includes 7500 satellites at 340 km altitude, which are made up of 150 orbital planes of 50 satellites each, with inclinations at 53.06°, and elevation angle at 40°. Without loss of generality, choices of the ISL transmission capacity, satellites' transmission power, and channel gain between satellites and mobile devices are set as $10 \sim 60$ Mbps, $50 \sim 150$ W, and $-170 \sim -180$ db, respectively. The total communication round K in FL is set to 20, and in every communication round, the number of ground mobile devices is taken as 1000, and the number of available satellites is set to $10 \sim 20$. For practical implementation of FL, we assume that the probability that the mobile devices attend the FL in each communication round is 0.1. (i.e., the number of devices attending the FL is about 100 in every round.) Also, we randomly initialize devices' server selection strategies and satellites' handover strategies. On the other hand, to highlight the heterogeneity of the mobile devices and satellites, the data size that every device owns D_m, their contribution index α_m, together with the satellites' transmission power E_n, follow a uniform distribution $U[a, b]$, where a and b are the upper and lower bounds, respectively (shown as $a \sim b$ in Table 6.1).

On the other hand, to evaluate the scalability robustness of the proposed MFEv approach, the number of ground devices is set to vary from 100 to 1000. Based on the satellite constellation, we classified the ground area into hexagonal grids, each with a radius of 1 km. For the purposes of this simulation, we considered that the ground devices are randomly located in the same grid, and the proposed MFEv approach is designed to be easily scalable to accommodate any number of grids through parallel computing.

The set of simulation parameters and their values are summarized in Table 6.1. It is worthwhile to mention that in practical SatComs, the satellites orbit in fully-predictable trajectories. Thus, the values of channel gain g_{m,n^k} between satellite n and device m, the handover indicator variable δ_{m,n^k}, the remaining parameter size

Table 6.1 Parameters for simulation

Parameters	Values
Total number of mobile devices M	1000
Devices' FL-attending probability	0.1
Number of available satellites N in every communication round	$10 \sim 20$
Devices' FL-attending probability	0.1
Devices' data size D_m	$100 \sim 500$ M
Devices' data contribution index α_m	$1.0 \sim 5.0$
Satellites' compensation bonus for devices $b_{m,n}^k$	$1 \sim 2$ \$
Satellites' transmission power in the ISL E_n	$50 \sim 150$ W
Satellites' handover indicator variable $\delta_{m,n}^k$	$\mathcal{B}(0.1)$
Channel gain between satellite n and device m, i.e., $g_{m,n}$	$-170 \sim -180$ db
Transmission capacity in the ISL $c_{n,n'}^k$	$10 \sim 60$ Mbps
Devices' payoff coefficient $\lambda_{m,n}^{pf}$	$3 \times 10^{-25} \sim 4 \times 10^{-25}$
Devices' communication cost coefficient $\lambda_{m,n}^{cm}$	$0.2 \sim 0.3$
Satellites' handover cost coefficient $\lambda_{m,n}^{hd}$	$5 \sim 5.5$
MFEv time step dt	10^{-3}
MFEv stopping threshold η_0	10^{-4}

$S_{m,n}^k$, and the transmission capacity in the ISL $c_{n,n'}^k$ between satellite n and n' are assumed to be known.

Remark 6.3 As illustrated in Remark 6.1, our work concentrates on server selection and handover optimization for the SatCom-based FL scheme, which falls within the realm of incentive mechanism design. Furthermore, the scheme we propose is adaptable to any FL model, data, and training strategies, without affecting the convergence of the FL. Given this focus, our evaluations center on the algorithm's convergence, performance, sensitivity, and scalability. We have opted not to undertake an FL training process in this context.

6.4.2 Convergence Analysis

We first illustrate the convergence analysis of the proposed MFEv approach. In Fig. 6.4, we randomly select 10 devices and show their convergence behavior. It can be seen that the proposed MFG approach converges fast. Specifically, as is shown in Fig. 6.4a, after about 60 iterations, mobile devices' overall cost converges and reaches the NE. In Fig. 6.4b, a reference devices' server selection strategy is shown. It can be seen that the device's strategy is initialized as a uniform probability, and converges fast within 60 iterations. Figure 6.4c, d, e depict devices' communication costs, payoffs, and handover costs, respectively. It is worthwhile to mention that in Fig. 6.4d, not all devices' payoffs increase with the iteration index. This is because

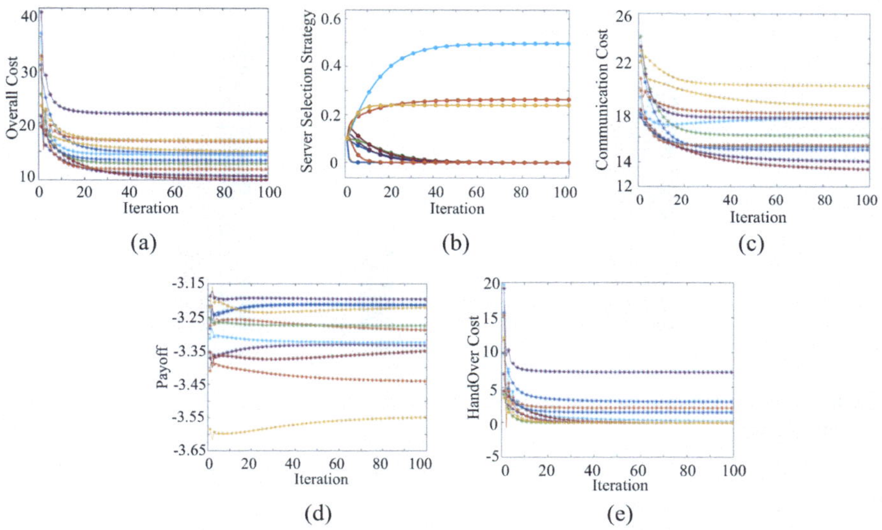

Fig. 6.4 Convergence of 100 devices' costs and strategies using the proposed MFEv approach (Only ten randomly selected devices' convergence behavior are plotted). (**a**) Convergence of overall cost. (**b**) Convergence of a reference device's strategy. (**c**) Convergence of communication cost. (**d**) Convergence of payoff. (**e**) Convergence of handover cost

some devices' communication costs make up a larger proportion of their utilities than their payoffs. For these devices, in order to minimize their overall costs, their communication costs reduce, which results in a decrease in their payoffs.

6.4.3 Performance Analysis

We first show how the devices' optimized server selection strategies and satellite handover strategies influence system performance. To be specific, we compare the costs of 1000 devices before and after the proposed MFEv optimization approach. As can be seen in Fig. 6.5, devices' overall costs are reduced by about 28% after the proposed MFEv optimization. More specifically, the devices' communication costs are reduced by about 10.5%, their payoffs increase by about 6%, and their handover costs decrease by about 82%, respectively. On the other hand, we compare and plot a reference device's server selection strategies before and after the proposed MFEv approach in Fig. 6.6. It can be seen that devices randomly select satellites initially. After the MFEv optimization, devices' server selection strategies focus on some specific satellites to minimize their costs.

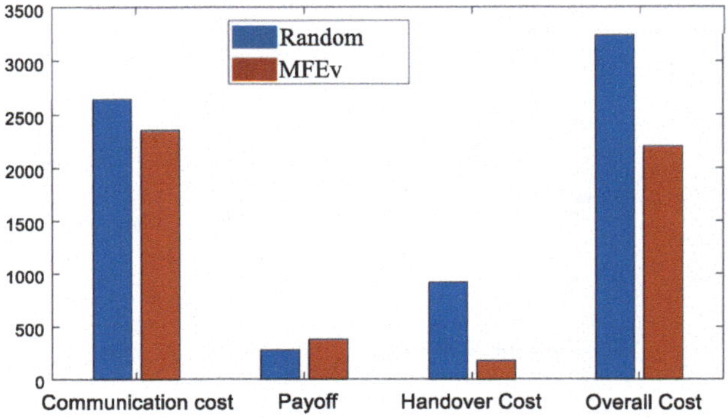

Fig. 6.5 The comparison of 100 devices' optimized costs between MFEv optimization and random strategy

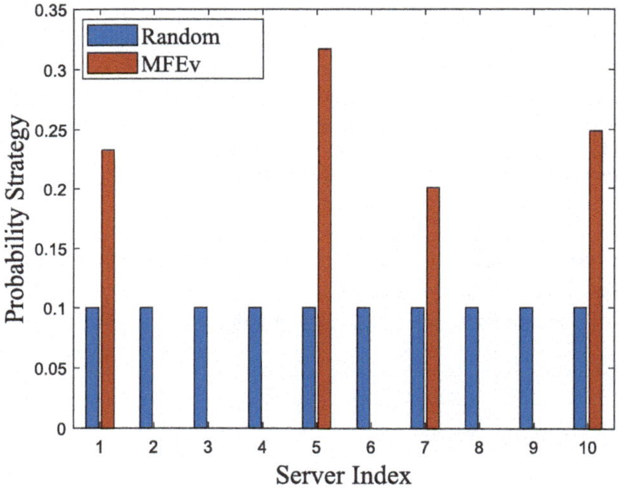

Fig. 6.6 The comparison of a reference device's server selection strategy between MFEv optimization and Random strategy

6.4.4 Sensitivity Analysis

We also demonstrate the sensitivity of the proposed MFEv approach. Specifically, from the convergence analysis in Sect. 6.3.5.1, we see that device communication has a great influence on devices' strategies. Thus, we look in these simulations into the sensitivity analysis of devices' averaged communication costs with-respect-to (w.r.t.) channel gain between satellites and devices in SatComs. Specifically, we randomly initialize channel gain g ranging from -170 db to -180 db. As can be seen in Fig. 6.7, devices' communication costs increase almost linearly as the channel gain decreases, which is consistent with common sense as lower channel

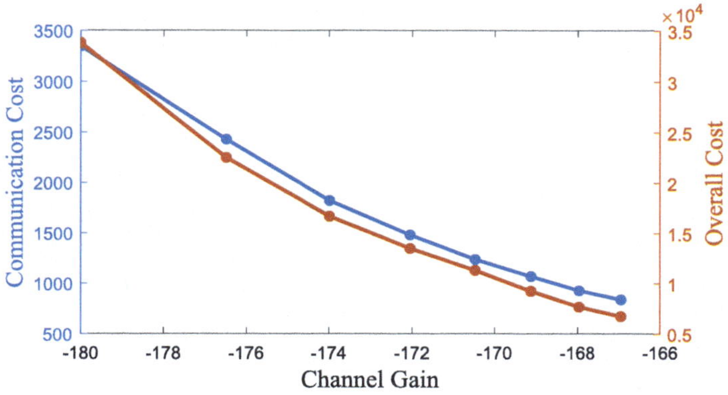

Fig. 6.7 Channel gain's influence of 100 devices' communication costs and overall costs

Table 6.2 Proportion of devices that selects satellites with handover

Total number of devices	Number of device choosing handover	Proportion
100	12	12%
200	18	9%
400	36	9%
600	65	10.8%
800	79	9.8%
1000	117	11.7%

gain means that devices need to consume more energy to transmit a comparable amount of data to satellites. On the other hand, we also investigate how satellites' handover costs influence devices' server selection strategies. As summarized in Table 6.2, the proportion of devices choosing satellites with handover service is very low, only $9.8 \sim 12\%$. That is, in order to reduce the handover cost, devices tend to select satellites that can receive all their model parameters within one communication round so that the satellite will not need handover service.

6.4.5 Comparison with State-of-the-Art Method

We finally analyze the scalability of the proposed MFEv method by comparing it with Random Strategy and a state-of-the-art incentive mechanism method—Replicator Dynamics (RD) approach [220], where the authors formulate an evolutionary game to model the dynamics of the cluster selection process and optimize with RD. As shown in Proposition 6.3, the time complexity of the proposed MFEv approach is $O(MN)$, where M is the number of mobile devices, and N is the number of satellites. We confirm this conclusion in the following comparative experiments, where all the implemented methods are coded in Matlab running on CPU: AMD

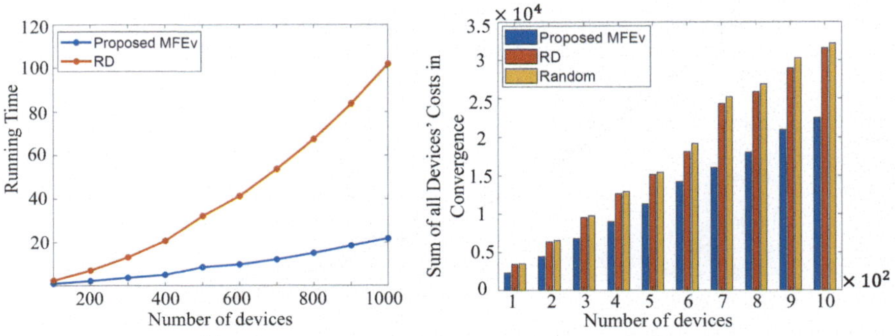

Fig. 6.8 Comparisons of 100 devices' costs and the code running times between our proposed MFEv, random strategy (Random), and replicator dynamics (RD) [220]. (**a**) Code running time comparison. (**b**) Devices' optimized cost comparison

Ryzen 7 5800X 8-Core and GPU: Nvidia Geforce RTX 3060ti. It can be seen in Fig. 6.8a that the code running time of the proposed MFEv approach is linear in the number of devices, and is much shorter than the RD approach. In fact, the code running time of the proposed MFEv is about 80% shorter than RD, confirming the efficiency of the proposed MFEv method. On the other hand, we also compared the optimized cost obtained by the two methods in Fig. 6.8b. It can be seen that compared with RD, our proposed MFEv can achieve much lower cost than RD and random strategy. Specifically, the cost optimized by the proposed MFEv is about 33% lower than the random strategy, and 31% lower than RD.

6.5 Chapter Summary

In this chapter, we have investigated the joint server selection and handover design problem for a large-scale SatComs-based FL framework. Interactions between a large number of devices will result in computational complexity for traditional games because they usually deal with each device separately. To address the issue of a large number of mobile devices, we have formulated the problem as a discrete MFG, and have proposed an MFEv algorithm to solve this problem. In the proposed MFEv algorithm, the behaviors of other devices are modeled as a mean-field which is a distribution over the strategy set. Theoretical analysis and simulation results have demonstrated that the proposed MFEv method can achieve the NE (the cost reduces by about 31%) while requiring a much lower computation complexity compared with SOTA methods (the code running time reduces by about 80%).

Chapter 7
Client Selection in Hierarchical Federated Learning with Mean Field Game

In this chapter, a client-edge-cloud hierarchical federated learning (FL) model has been developed for driving range estimation (DRE) of battery electric vehicles (BEV). Generalized models are aggregated on the cloud server, while customized models trained on local data with similar data distribution are aggregated on the edge server, which mitigates the impact of data heterogeneity. To reduce the communication overhead of FL, clients will periodically update to the edge server and edge servers will periodically update to the cloud server. Moreover, we propose a mean field game-based probabilistic device selection scheme. Jointly considering their contributions and the population diversity, a fraction of devices will be selected to join the FL iteration.

7.1 Hierarchical Federated Learning Framework

The hierarchical FL systems have three layers as shown in Fig. 7.1. In the client layer, BEVs are running in a variety of locations with different temperatures and geography. Vehicles are organically clustered by location and are linked to local edge servers, which periodically update their aggregated model parameters to a remote central cloud server. (d, v, s) are the indexes of data, vehicle, and server, respectively.

7.1.1 Local Model Training on Vehicles

Denote the training dataset on a generic vehicle as $\mathcal{D} = \{x_d, y_d\}_{d=1}^{|\mathcal{D}|}$, where $|\mathcal{D}|$ refers to the total number of training samples. The input data x_d includes the average speed, acceleration, jerk of acceleration, initial battery SoC, temperature

Y. Kang et al., *Mean Field Guided Machine Learning*, Wireless Networks,
https://doi.org/10.1007/978-3-031-91859-9_7

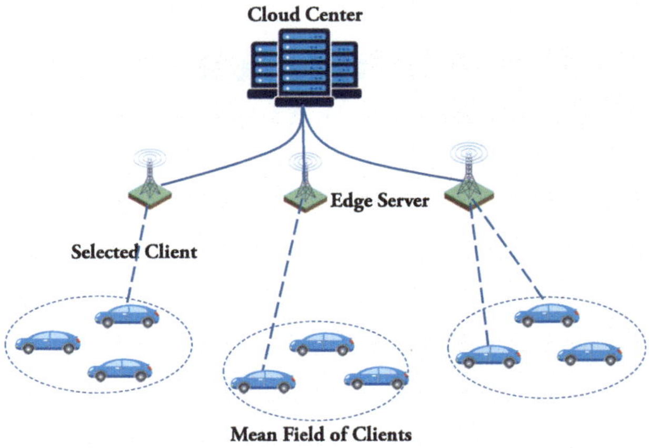

Fig. 7.1 An illustration of the client-edge-cloud hierarchical federated learning structures

,and geography conditions (plain or hill), while the label y_d is the remaining driving range. Parameters of the local ML model are denoted as ω_v and the loss function of $f_d(x_d, y_d, \omega)$ refers to the loss of data sample (x_d, y_d). The training process on a generic vehicle is to minimize the local empirical loss $F(\omega)$ given the training dataset $\mathcal{D}$

$$F(\omega) = \frac{1}{|\mathcal{D}|}\sum_{d=1}^{|\mathcal{D}|} f(x_d, y_d, \omega) = \frac{1}{|\mathcal{D}|}\sum_{d=1}^{|\mathcal{D}|} f_d(\omega). \tag{7.1}$$

Without loss of generality, we assume that the local model parameter ω is updated by gradient descent

$$\omega(k) = \omega(k-1) - \alpha \nabla F(\omega(k-1)), \tag{7.2}$$

where the update step is denoted by k and the learning rate is α. The vehicle index v is dropped in (7.1) and (7.2), as we only consider a generic vehicle in this subsection.

7.1.2 Periodical Model Aggregation on Servers

Frequent FL parameter uploading from vehicles to edge servers and cloud servers causes a significant delay, which can be orders of magnitude more than the training time of ML model [146]. To efficiently leverage the communication resources, vehicles will update their model parameters to the edge server every p_1 FL iteration. Edge servers update their aggregated model parameters to the cloud sever every $p_1 p_2$ FL iteration, i.e., the model on the central cloud server will update one time when the model on the edge server updates p_2 times. The empirical loss function of edge server s is given by

$$F_s(\omega) = \frac{1}{|\mathcal{D}^s|} \sum_{v \in \mathcal{A}_s} |\mathcal{D}_v^s| F_v(\omega), \tag{7.3}$$

where selected vehicles connected to the edge server s are denoted by the set $\mathcal{A}_s = \{\mathcal{D}_v\}_{v=1}^{|\mathcal{A}_s|}$, the vehicle v is represented by its dataset $\mathcal{D}_v$, and $|\mathcal{D}^s| = \sum_{v \in \mathcal{A}_s} |\mathcal{D}_v^s|$ is the total number of training samples on the vehicles connected to edge server s. The model parameters on edge server s are updated by

$$\begin{aligned} \omega^s(k) &= \frac{1}{|\mathcal{D}^s|} \sum_{v \in \mathcal{A}_s} |\mathcal{D}_v^s| \omega_v^s(k), \\ &= \frac{\sum_{v \in \mathcal{A}_s} |\mathcal{D}_v^s| [\omega_v^s(k-1) - \alpha_v^s \nabla F_v(\omega_v^s(k-1))]}{|\mathcal{D}^s|}, \end{aligned} \tag{7.4}$$

where $k|p_1 = 0$ and α_v^s is the learning rate of vehicle v. Similarly, the update of the global model on the cloud server is given by

$$\begin{aligned} \omega^g(k) &= \frac{\sum_{v=1}^N |\mathcal{D}_v| \omega_v(k)}{|\mathcal{D}_g|}, \\ &= \frac{\sum_{v=1}^N |\mathcal{D}_v| [\omega_v(k-1) - \alpha_v \nabla F_v(\omega_v(k-1))]}{|\mathcal{D}_g|}, \end{aligned} \tag{7.5}$$

where N is the total number of vehicles in the FL system, $k|p_1 p_2 = 0$, $|\mathcal{D}_g| = \sum_{v=1}^N |\mathcal{D}_v|$ denotes the total number of vehicles in the system, and ω^g represents the parameters of the global model.

7.2 Mean Field Game on Graph for Probabilistic Device Selection

The purpose of device selection is to reduce the communication overhead by selecting part of the devices to participate in the FL iteration. How to determine the selection probability of each device is the key to reducing the delay and accelerating the training.

7.2.1 Vehicle to Vehicle Communication Network

As shown in Fig. 7.2, we consider a V2V communication network consisting of several nearby vehicles. Two vehicles are connected and share information only if they trust each other. The contribution of a vehicle is determined by many empirical factors, such as the norm of the gradient, the distance to the edge server [33], and

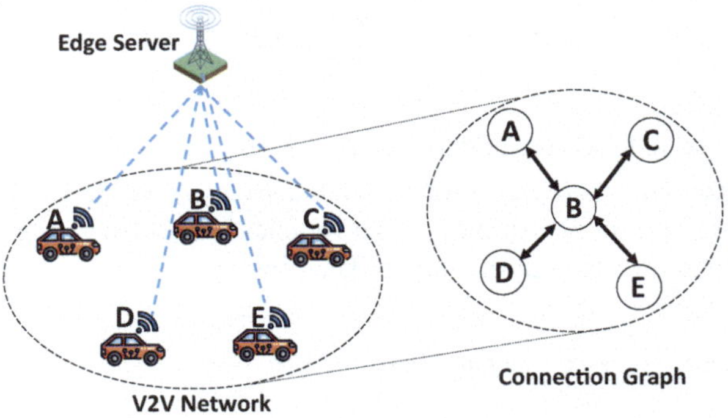

Fig. 7.2 An example of V2V communication networks

vehicle data similarity [190]. A larger gradient typically leads to faster convergence of the training, while a closer distance to the edge server normally results in a shorter communication delay. As for data similarity, a completely different dataset can better generalize the ML model. In our methodology, the contribution is only determined by the gradient. However, the proposed probabilistic device selection scheme can be adapted in cases where the contributions are quantized by many factors, because the computation of device selection probability is only related to the value of contribution and independent of how it is computed. The contribution of a generic vehicle v in p_1 SGD iterations (one FL iteration) is given by

$$c_v = ||\sum_{k=1}^{p_1} \sum_{d \in \mathcal{D}_v} \nabla f(\omega_v(k-1), x_{v,d}, y_{v,d})||, \tag{7.6}$$

where the index of the connected server is dropped as we now consider vehicles connected to the same edge server and $||\cdot||$ denotes the Euclidean norm.

7.2.2 Mean Field Game on Graph

Mean field game (MFG), first developed by Lasry and Lion [117], is a general game-theoretic framework for situations involving a large number of agents [70, 71]. MFG can describe the evolution of the mean field (continuous or discrete probability distribution of state variable) by a partial differential equation (PDE) called the Fokker-Planck-Kolomogorov (FPK) equation, while satisfying certain optimal conditions represented by another PDE named the Hamilton-Jacobi-Bellman equation (HJB). *The device selection probability distribution is the mean field and its evolution can be modeled by the FPK equation.*

We consider a population of u vehicles that are connected to a generic edge server, denoted by a set $\mathcal{B} = \{1, 2, \cdots, u\}$. Each vehicle is regarded as a node on the graph $\mathcal{G} = (\mathcal{B}, \mathcal{E})$ and two nodes $i, j \in \mathcal{B}$ are connected by an edge $(i, j) \in \mathcal{E}$ if two vehicles can share information. In MFG, the vehicles are discrete strategies. The target is to find the selection probability distribution (mean field), which is a vector denoted by $\rho = \{\rho_v\}_{i=1}^{u}$ and ρ_v is the probability of selecting vehicle v. ρ is often known as the mixed strategy, which belongs to the simplex

$$\mathcal{P}(\mathcal{B}) = \left\{ (\rho_v)_{v=1}^{u} \in \mathbb{R}^u : \sum_{v=1}^{u} \rho_v = 1,\ 0 \leqslant \rho_v \leqslant 1, v \in \mathcal{B} \right\}. \tag{7.7}$$

Assigning the selection probability proportional to the contributions of vehicles is a common and simple idea [33]. However, there are two major problems with this idea: (1) the contribution information of all vehicles must be uploaded to the edge server and the computation of selection probability is centralized, which may cause some privacy issues (2) some low-contribution vehicles might be completely ignored even though they have important data. To overcome the above two drawbacks, we first define a utility function

$$\mathcal{J}(\rho) = \sum_{v=1}^{u} c_v \rho_v - \beta \sum_{v=1}^{u} \rho_v \log \rho_v, \tag{7.8}$$

which consists of the weighted contribution and the Shannon-Boltzman entropy. To maximize the weighted contribution, we need to assign more weight to the high-contribution vehicle (exploitation). To maximize the Shannon entropy, we need to assign the weight more equally to all vehicles (exploration). β is a hyperparameter making the trade-off between exploration and exploitation.

To maximize the utility function in (7.8), we need to introduce the discrete Fokker-Planck equation as follows:

Theorem 7.1 *Given a discrete strategy graph $\mathcal{G} = (\mathcal{B}, \mathcal{E})$, the payoff function to strategy i $\mathcal{F}_i : \mathcal{P}(\mathcal{B}) \to \mathbb{R}$ is continuous. Then for any initial condition $\rho_0 \in \mathcal{P}(\mathcal{B})$, the Fokker-Planck equation*

$$\frac{\mathrm{d}\rho_i}{\mathrm{d}t} = \sum_{j \in N(i)} \frac{1}{d_j} \rho_j [\mathcal{F}_i(\rho) - \mathcal{F}_j(\rho)]^+ - \sum_{j \in N(i)} \frac{1}{d_i} \rho_i [\mathcal{F}_j(\rho) - \mathcal{F}_i(\rho))]^+ \tag{7.9}$$

is a well-defined gradient flow in $\mathcal{P}(\mathcal{B})$. $[\cdot]^+ = max\{\cdot, 0\}$, $N(i)$ is the neighborhood of node i, and the degree of node j is denoted as $d_j = \sum_{i \in N(j)} 1$.

The theoretical proof of Theorem 7.1 is given in [39]. For the device selection problem in the FL, the discrete strategy is the vehicle and the payoff function of selecting vehicle v is computed by

$$\mathcal{F}_v(\rho) = \frac{\partial \mathcal{J}}{\partial \rho_v} = c_v - \beta \log \rho_v. \tag{7.10}$$

Updating the selection probability with (7.9), vehicles only need to send their contribution information to their trusted neighbors and the computation of selection probability is completely decentralized via the V2V network.

7.3 Performance Evaluation

7.3.1 *Hierarchical Structure*

We consider a 3-layer hierarchical FL structure in the experiment. In the client layer, there are 100 intelligent vehicles evenly located in 4 different cities, i.e., there are 25 vehicles that will join the FL in each city. These cities vary in temperature and topography as shown in Table 7.1. In each city, there is an edge server serving the vehicles in the same city, thus the aggregated model on the edge server is customized for users in the corresponding city. The edge servers across cities do not share model parameters. In the cloud layer, there is only one central cloud server, which will aggregate the information from the 4 edge servers to obtain a global model. The vehicles will update to the corresponding edge server every $p_1 = 5$ FL iterations and the edge server will update to the cloud server every $p_1 p_2 = 10$ FL iterations.

7.3.2 *Dataset and Training*

With Aimsun, we collected the simulated data of more than 200 vehicles in each city. Each vehicle will generate a 60-min dataset consisting of critical information related to DRE, such as the average speed, the average acceleration, the current SoC, the average jerk, temperature, and topography.

For the local training, we build a basic long short-term memory (LSTM) network consisting of two linear layers and one LSTM layer with PyTorch. The input size, the feature embedding size, and the hidden size of LSTM are 56, 128, and 256, respectively. There are 10 local epochs on the vehicle during each FL iteration, with the learning rate of 0.001 and batch size 16. We use a stochastic gradient descent optimizer with momentum 0.5. For the probabilistic device selection, we set the regularization term $\beta = 0.3$.

Table 7.1 Configuration of 4 cities

Notation of city	A	B	C	D
Temperature (°C)	0	0	20	20
Topography	Hilly	Flat	Hilly	Flat

7.3.3 Significance of Customization with Hierarchy

In Fig. 7.3, we highlight the importance of customization by comparing the performance of several models on a dataset of randomly selected users in city A. The customized model is the model that was aggregated on the edge server in city A. Other models are those that have been adapted for users other than those in city A. The general model is the global model, which is identical to the global model in FedAvg [18]. When the performance of these models is compared, we see that the customized model for city A outperforms the global model, which outperforms the models customized for other cities. In Fig. 7.4, the significance of customization is proved by comparing the prediction results of the model customized for city C on the test datasets obtained from all cities. The model performs best on the dataset from city C and worst on the dataset from city A, demonstrating that the model's performance is significantly dependent on data heterogeneity between training and test data. The hierarchical structure splits users into different groups and reduces data heterogeneity within the same group. The data heterogeneity within the same

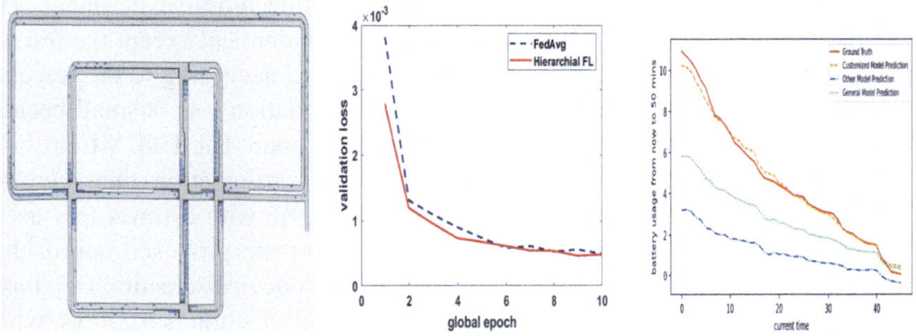

(a) (b) (c)

Fig. 7.3 (**a**) Simulation environment in Aimsun. (**b**) Convergence comparison between hierarchical FL and FedAvg. (**c**) SoC prediction results of different models

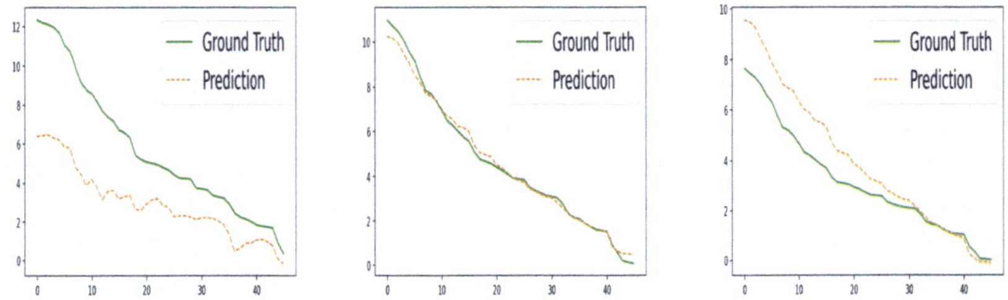

(a) (b) (c)

Fig. 7.4 We use the model customized for city C to predict the driving range of vehicles in all cities. (**a**), (**b**), and (**c**) depicts the predictions on the dataset collected in city A, C, and D, respectively

Table 7.2 Comparison on the validation loss

Model	Fraction of users	Validation loss
FedAvg [18]	%100	0.00121
HierFed V1	%10	0.00124
HierFed V2	%20	0.00086

group is decreased since these cars gather data from the same environment, which allows for the customized model's great performance, as seen by the orange curve in Figs. 7.3c and 7.4b.

7.3.4 Comparison with Baselines

We evaluate the performance of the model by computing the mean square error (MSE) of the continuous SoC predictions, as shown in Fig. 7.3c. The proposed hierarchical FL achieves similar convergence speed to the FedAvg with 10 times fewer users' participation, as shown in Fig. 7.3b. Table 7.2 compares our proposed approach with the FedAvg, which serves as the baseline in our experiment. Two hierarchical models (HierFed V1 and HierFed V2) are identical except the fraction of users that participate in the FL and users are selected according to the selection scheme based on MFG. The absolute value of the validation loss is small because we are considering the change of SoC within one hour. HierFed V1 provides comparable performance with 10 times less users' participation than FedAvg, and HierFed V2 improves prediction accuracy by 28.9% with 5 times less users' participation than FedAvg. The computational time of the proposed approach is similar to that of FedAvg, even though there is an extra device selection step based on MFG. The average computational time of the device selection is 0.2775 s, which is negligible compared with the training of the LSTM model.

7.4 Chapter Summary

In this chapter, we propose a hierarchical federated learning framework for connected vehicle applications and take driving range estimation of battery electric vehicles as an example. We use two strategies to decrease communication overhead: periodic averaging and MFG-based probabilistic device selection. The mean field game-based device selection scheme is completely decentralized and vehicles only need to share minimal information with trustworthy neighbors in the local V2V networks, which enhances the users' privacy. The client-edge-cloud structure may efficiently divide users into distinct groups based on their geographic locations, reducing data heterogeneity within the same group. As a result, the model on the edge server is customized for users in the specific region, increasing the model accuracy by 28.9% with 5 times less users' participation than the vanilla federated learning.

Chapter 8
Evolutionary Neural Architecture Search with Mean Field Game Selection Mechanism

Deep Neural Networks (DNNs) have found great success in a wide range of applications, including image classification, speech recognition, and natural language modeling. In general, the performance of DNNs is determined by two factors: their architectures and the weights associated with them. Weights are usually optimized through a learning process, which involves using a continuous loss function to measure the discrepancies between the prediction and the real one, and then using gradient-based algorithms to minimize the discrepancies. Promising architectures of DNNs are often manually designed with rich expertise and knowledge in the specific task. Nowadays, due to the emergence of the Internet-of-Things, datasets grow significantly larger and are collected locally in an unbalanced and none i.i.d. fashion. Given that the DNN architectures are often data-dependent, the edge users need to redesign the architectures accordingly. However, they usually lack such kind of expertise and knowledge, which raises the need for an automatic neural architecture design.

Neural Architecture Search (NAS), which aims to automate the DNN architecture design, is a promising technique to handle the aforementioned challenge. Existing NAS algorithms can be broadly classified into three categories based on the optimizer used: Reinforcement Learning (RL) based NAS algorithms, gradient-based NAS algorithms, and Evolutionary Computation (EC) based NAS algorithms (ENAS). RL based NAS often requires a relatively large amount of computational resources even on median-scale datasets. Gradient-based NAS are more efficient but frequently discover ill-conditioned architectures due to incorrect relations for adapting to gradient-based optimization. ENAS algorithms search the optimal architectures through EC, which is a class of population-based algorithms simulating the evolution of organisms and species in nature. They are well-suited for non-convex, non-linear, or non-smooth optimization problems. With no assumptions on the optimization landscape, evolutionary NAS can evolve the DNNs towards better architectures based on genetic variation and selection.

Y. Kang et al., *Mean Field Guided Machine Learning*, Wireless Networks,
https://doi.org/10.1007/978-3-031-91859-9_8

Evolutionary NAS can be generally regarded as a combination of two problems: *weight optimization* and *evolutionary architecture search*. Most of the literature focus on the first problem. In general, they put significant efforts into developing more efficient weight optimization techniques to reduce the computational cost of evolutionary NAS. Weight sharing approaches are the mainstream of those research. However, from the perspective of evolution, all these approaches only address one issue in the evolutionary process—*fitness evaluation*. Another unattended key issue of the evolutionary NAS is *what kind of architecture should be preserved and passed through the generations*. In other words, how can we *select parents* from each generation so that the finite number of representative agents can more extensively search the architecture space and find the optimal architecture? In this regard, we first identify the selection problem as a *mean field game* (MFG), where the selection probability distribution on the mating candidates is the mean field information that needs to be dynamically controlled. The utility function is defined with the target of maximizing the *average fitness* and the *population diversity*. Based on the MFG formulation, it is evident that neither uniform random selection nor roulette wheel selection [46], which are frequently used, is the optimal selection strategy. Furthermore, we use the discrete Fokker-Planck equation to find the optimal solution to the proposed MFG and use the derived selection probability to select parents in each generation.

8.1 Evolutionary Neural Architecture Search

The NAS problem is decoupled into two sequential steps: weights optimization and evolutionary architecture search, as shown in Fig. 8.1. Formally, the first step is the supernet weight optimization given by

$$W_{\mathcal{S}} = \arg\min_{W} \mathcal{L}_{\text{train}}(\mathcal{N}(\mathcal{S}, W)), \tag{8.1}$$

where W is the weight of the supernet, $\mathcal{S}$ is the architecture search space, $\mathcal{L}$ is the loss function, and $\mathcal{N}$ denotes the neural network. The second step is architecture search given by

$$a^* = \arg\max_{a \in \mathcal{S}} V(\mathcal{N}(a, W_{\mathcal{S}}(a))), \tag{8.2}$$

where V denotes the validation accuracy on the given dataset and a refers to a sampled architecture from the supernet. Any sampled architecture can be regarded as a single path in the supernet as shown in Fig. 8.1. The weights $W_{\mathcal{S}}(a)$ of the sampled neural network are directly inherited from the supernet, which makes the search *efficient* and *affordable* even with large datasets, such as ImageNet.

The rest of this section is organized as follows: In Sect. 8.1.1, we discuss the details of the weight optimization of the supernet. In Sect. 8.1.2, we introduce the

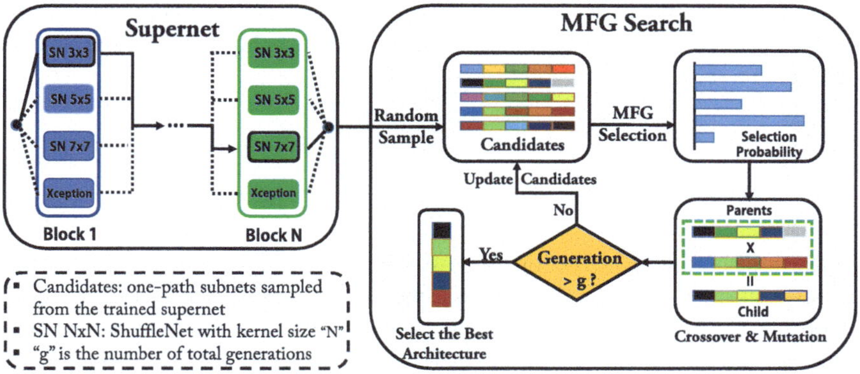

Fig. 8.1 Overall framework of the single path one-shot neural architecture search with MFG selection. The supernet consists of N different choice blocks and each choice block contains ShuffleNet choices with different kernel sizes. Each path through all choice blocks is one candidate neural architecture. In the search stage, several candidates will be sampled from the supernet and inherited the weights. After fitness evaluation, their selection probability will be computed via MFG. Selected parents will generate the offspring through crossover and mutation. The population keeps evolving until the final generation

overall evolutionary process. In Sect. 8.2, we formulate the selection problem as an MFG and propose the Fokker-Planck equation to solve it.

8.1.1 Supernet Weight Optimization

Like other weight sharing approaches, the core target of training the supernet is that, the validation accuracy of any sampled architecture a using inherited weight $W_{\mathcal{S}}(a)$ is a highly predictive accuracy of the fine-tuned architecture a. This leads to the training principle that the supernet weights $W_{\mathcal{S}}$ should be optimized in a way that all architectures in the search space are optimized simultaneously [82]. In this regard, the supernet is regarded as a *stochastic* neural network, which can represent different architectures given a prior distribution $\Psi(\mathcal{S})$. Formally, weight optimization is described by

$$W_{\mathcal{S}} = \arg\min_{W} \mathbb{E}_{a\in\Psi(\mathcal{S})}\{\mathcal{L}_{\text{train}}(\mathcal{N}(\mathcal{S}, W(a)))\}. \tag{8.3}$$

The supernet consists of multiple choice blocks as shown in Fig. 8.1 and each choice block contain several choices related to the neural network architecture, such as the channel number. Each time we will sample a combination of choices based on the prior distribution $\Psi(\mathcal{S})$. The distribution $\Psi(\mathcal{S})$ can significantly influence the NAS algorithm and we adopt a uniform distribution in this chapter, which is an empirically good option. Furthermore, the distribution will be fixed during the training process. Every sampled architecture is a *single path* that connects architectures of different choice blocks, which can reduce the co-adaptation between node weights [82].

8.1.2 Evolutionary Algorithm

The starting point of the evolution process is the initialization of populations, which are different architectures sampled from the supernet as shown in Fig. 8.1. Each architecture is represented by a tuple and each entry of the tuple denotes the choice of the corresponding block in the supernet. All the candidate architectures will inherit their weights from the supernet. The validation accuracy is regarded as the *fitness* of the candidate architecture and is obtained by forward passing the data through the architecture. Thanks to the pre-training of the supernet, there is no need for backpropagation and weight optimization.

There are two steps in the selection process. In the first step, we use the *elitist* strategy [16, 46] to select the mating candidates, where u individuals with the best fitness are first selected from the population. The unselected networks are then removed from the population–it is *killed*. In the second step, we analytically compute the selection probability of the mating candidates, where we jointly consider the average fitness and population diversity of the selected parents. This step will be discussed later in detail. According to the selection probability, we will select *parents* from the mating candidates and they will reproduce *child*. Each *gene* of the child, which is a number denoting the choice from the corresponding choice block, is randomly copied from *father* or *mother*. At last, we apply *mutation* to each gene of the child, where we will mutate the gene with a certain probability. The probability is usually small because we want to randomly introduce some new features to the networks instead of changing the direction of evolution completely.

8.2 MFG Probabilistic Selection

In the first step of the selection process, we use the elitist strategy to select u mating candidates from the current generation. In the existing literature, researchers usually randomly select parents or select parents proportional to their fitness [46] (Roulette Wheel Selection [77]). However, given that the architecture search space is a complex and multimodal domain, the aforementioned two selection strategies have some evident drawbacks. If we randomly select parents from the mating candidates, we cannot ensure the best average fitness of the parents because a network with lower fitness has an equal probability to be selected. If we select parents proportional to their fitness, we get better average fitness of the parents. However, some simple architectures which lead to higher fitness at the early stage of the evolution process might readily dominate the population limiting the population diversity. In other words, the evolutionary algorithm will quickly converge to a single peak (local optima) in the search space, which is called *genetic drift* [49]. In this section, we analytically compute the selection probability of each mating candidate with MFG, which aims at ensuring average fitness and maintaining population diversity.

8.2.1 MFG Formulation

We consider a population of u mating candidates, denoted by a set $\mathcal{M} = \{1, 2, \cdots, u\}$. In the MFG, the u mating candidates are regarded as discrete pure strategies. The selection probability distribution (mean field) is a vector denoted by $\rho = \{\rho_i\}_{i=1}^{u}$, where ρ_i is the probability of selecting candidate i as one of the parents. ρ is often known as mixed strategy, which belongs to the simplex as follows:

$$\mathcal{P}(\mathcal{M}) = \left\{ (\rho_i)_{i=1}^{u} \in \mathbb{R}^u : \sum_{i=1}^{u} \rho_i = 1,\ 0 \leqslant \rho_i \leqslant 1, i \in \mathcal{M} \right\}. \tag{8.4}$$

The fitness of mating candidate i is denoted by f_i. The average fitness of the selected parents is given by

$$\mathcal{A}(\rho) = \sum_{i=1}^{u} f_i \rho_i. \tag{8.5}$$

If we ignore the population diversity, then the solution to (8.5) is evident and trivial. We can just let the fittest candidate reproduce and replace other candidates in the population. However, without good population diversity, we cannot explore the multimodal domain extensively. In biological science, the Shannon entropy is a standard measure of population diversity [171] and has been used to measure structural heterogeneity for a long time [144]. Based on Shannon entropy, the population diversity of the selected parents can be given by

$$\mathcal{D}(\rho) = -\beta \sum_{j=1}^{u} \rho_i \log \rho_i, \tag{8.6}$$

where f_i is the fitness of candidate i and $\beta \geqslant 0$ is a constant. It is well-known that $\mathcal{D}$ achieves the maximum value with $\rho_i = 1/u, \forall i \in \mathcal{M}$, which will reduce the MFG selection to a uniform random selection. However, it will significantly reduce the average fitness. In other words, bad architectures might be preserved across the evolutionary process. Jointly considering the average fitness in (8.5) and population diversity in (8.6), we formulate the selection problem as an MFG as follows:

$$\max_{\rho} \ \mathcal{J}(\rho) = \mathcal{A}(\rho) + \mathcal{D}(\rho) = \sum_{i=1}^{u} f_i \rho_i - \beta \sum_{j=1}^{u} \rho_i \log \rho_i \tag{8.7}$$

subject to $\rho \in \mathcal{P}(\mathcal{M})$.

8.2.2 Fokker-Planck Equation

Equation (8.7) is concave because the average fitness $\mathcal{A}$ is affine and Shannon entropy $\mathcal{D}$ is concave. Therefore, we can find an optimal point in the simplex $\mathcal{P}(\mathcal{M})$ to maximize the joint utility $\mathcal{J}$. To solve (8.7), we first introduce Theorem 8.1 as follows.

Theorem 8.1 *Given a discrete strategy graph $\mathcal{G} = (\mathcal{M}, \mathcal{E})$, the payoff function to strategy i $F_i : \mathcal{P}(\mathcal{M}) \to \mathbb{R}$ is continuous. Then for any initial condition $\rho_0 \in \mathcal{P}(\mathcal{M})$, the Fokker-Planck equation*

$$\begin{aligned} \frac{\mathrm{d}\rho_i}{\mathrm{d}t} &= \sum_{j \in N(i)} \frac{1}{d_j} \rho_j [F_i(\rho) - F_j(\rho)]^+ \\ &- \sum_{j \in N(i)} \frac{1}{d_i} \rho_i [F_j(\rho) - F_i(\rho))]^+ \end{aligned} \tag{8.8}$$

is a well-defined gradient flow in $\mathcal{P}(\mathcal{M})$. β represents the strength of Shannon entropy, $[\cdot]^+ = max\{\cdot, 0\}$, $N(i)$ is the neighborhood of node i, and the degree of node j is denoted as $d_j = \sum_{i \in N(j)} 1$.

Theoretical proof of Theorem 8.1 is provided in [39]. The discrete strategy graph $\mathcal{G} = (\mathcal{M}, \mathcal{E})$ is constructed based on the discrete strategy set $\mathcal{M}$, i.e., the set of mating candidates. Each mating candidate is regarded as a node on the graph $\mathcal{G}$. Two nodes i and j form an edge $(i, j) \in \mathcal{E}$ if they are connected. The neighborhood $N(i)$ of node i is a set of nodes that are connected with node i. The strategy graph $\mathcal{G}$ of the selection problem in (8.7) is a *complete graph* because all mating candidates are equally connected. The potential function F_i of candidate i is computed by

$$F_i(\rho) = \frac{\partial \mathcal{J}}{\partial \rho_i} = f_i - \beta \log \rho_i. \tag{8.9}$$

F_i refers to the increase of utility by increasing the selection probability ρ_i given the current selection probability distribution ρ. In other words, if we increase the selection probability of candidate i, how much can we increase the population diversity and average fitness? We compare candidate i with every other candidate j based on their potential of increasing the global utility and we will move the

Algorithm 6 Evolutionary neural architecture search with MFG selection

Input: generations T, population size p, search space S, mating candidates size u, step size η

Train the supernet based on (8.3) with a uniform prior distribution $\Psi(\mathcal{S})$
Sample the initial population of architectures from supernet $\Gamma_p = \{a_1, a_2, \cdots, a_p\}$
for $t = 1 : T$ **do**
  for $i = 1 : p$ **do**
    // in parallel
    Forward pass the data through the neural network and obtain its fitness
  end for
  // evolution starts
  sort the candidates based on their fitness and select the mating candidates $\mathcal{M} = \{1, 2, \cdots, u\}$
  // compute the selection probability ρ
  randomly initialize ρ
  while not converge **do**
    for $i = 1 : u$ **do**
      compute the gradient $\frac{\mathrm{d}\rho_i}{\mathrm{d}t}$ with (8.8)
      update selection probability of candidate i: $\rho_i := \rho_i + \eta\frac{\mathrm{d}\rho_i}{\mathrm{d}t}$
    end for
  end while
  select the parents from mating candidates $\mathcal{M}$ with selection probability ρ
  reproduce a child with parents and add the child to the population until the size reaches p
end for
return networks with the best performance

selection probability from lower potential candidate to higher potential candidate with a switching rate proportional to the difference between their potentials.

In conclusion, the discrete Fokker-Planck equation in (8.8) tells us how to adjust the selection probability of each candidate i given the current selection probability distribution ρ. We can find the optimal selection probability distribution ρ by updating the selection probability iteratively based on (8.8). The evolutionary neural architecture search with MFG selection is summarized in Algorithm 6.

8.3 Experimental Results

Dataset and Training We conduct the experiments on ImageNet [161]. The original training set is randomly split into two parts: 50,000 images, organized in 1000 categories are used for validation and the rest are used for training. The original validation set is used for testing. We use the same experimental settings as [82], including data augmentation and learning schedule. In particular, for the training of supernet, the batch size is 256, the total iterations are 150,000, the initial learning rate is 0.5 and the optimizer is stochastic gradient descent with a momentum of 0.9 and weight decay of 4×10^{-5}. We use 2 NVIDIA Tesla P100 GPUs for training.

Table 8.1 Configuration of blocks

Choices	#Layer	Structure	Kernel
ShuffleNet	3	pw-dw-pw	3x3
ShuffleNet	3	pw-dw-pw	5x5
ShuffleNet	3	pw-dw-pw	7x7
Xception	6	dw-pw-dw-pw-dw-pw	3x3

Table 8.2 Configuration of supernet

Blocks	Input	#Repeat	#Channel
CB1	$112^2 \times 16$	4	64
CB2	$56^2 \times 64$	4	160
CB3	$28^2 \times 160$	8	320
CB4	$14^2 \times 320$	4	640

The population diversity β and the mutation rate μ are two important hyperparameters that can influence the MFG search algorithm significantly. Grid search is used to find the best combination of them. The maximum and minimum values of these two hyperparameters are empirically set to 0.3 and 0.1, respectively. We choose the grid size 0.05 for both parameters and every position in the grid is evaluated, which means the total number of combinations that have been verified is 25.

Search Space The basic choice block is constructed based on the state-of-the-art manually-designed networks: ShuffleNet v2 [141]. As shown in Table 8.1, each choice block consists of three different ShuffleNet v2 with three different kernel sizes and one ShuffleNet Xception. Convolutional layers include *depth-wise (dw)* layers and *point-wise(pw)* layers. The order of connection is shown in Table 8.1. The supernet consists of four types of choice blocks (CB) as shown in Table 8.2 and they vary in the input sizes and the number of channels. Each block will repeat several times making a supernet with 20 blocks. Therefore, the size of the search space is 4^{20}. Each sampled neural network is denoted by a tuple with 20 dimensions, where each dimension records the choice from the corresponding choice block.

Evolutionary Search Analysis The population size of each generation is $p = 50$ and the top $u = 10$ networks with better validation accuracy will be selected as mating candidates. We use the discrete Fokker-Planck equation in (8.8) to compute the selection probability for each mating candidate. The selection probability distribution on the mating candidates is recorded in a vector and a *gradient matrix* is constructed based on (8.8). The selection probability computation is very efficient and the average computation time of the probability computation is 0.1780 s for every generation, which is negligible compared with the training and evaluation of DNNs. The total iterations of the MFG are set to 1000 and the step size for updating the selection probability is $\eta = 0.01$. The mating candidates are selected according to the selection probability. For each reproduction, a *male* network and a *female* network will be selected. Their child will randomly inherit the choice from one of the parents for each block.

In Fig. 8.4a, we compare the evolutionary search algorithm with different selection methods, including proportional selection (roulette wheel selection), uniform selection, and MFG-based selection. Random search is also depicted for comparison. The search is conducted on the same supernet, with network design shown in Tables 8.1 and 8.2. A random seed is used to sample networks for the first generation so that all search algorithms have the same starting point. Based on the convergence analysis, the total number of generations of evolution is set to 15, as the evolutionary algorithm can empirically converge around 15 generations shown in Fig. 8.2a. The search with MFG selection achieves the best performance in terms of the top-1 validation error.

In Fig. 8.3a, we show how the mutation rate affects the search method. During this experiment, three unique mutation rates, which indicate the chance of picking

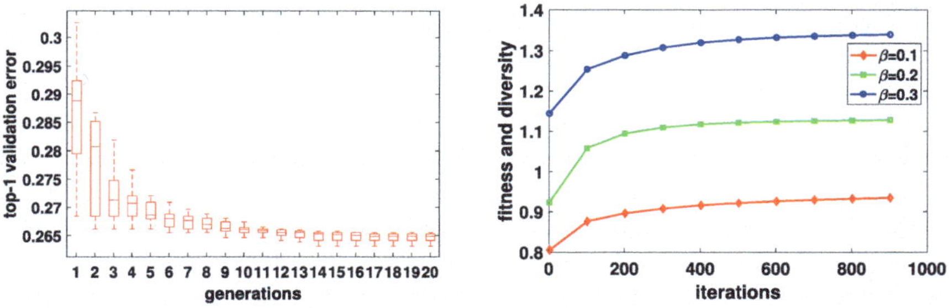

(a) Evolutionary Search Algorithm (b) MFG Computation

Fig. 8.2 Convergence Analysis. (**a**) The boxplots depict the distribution of validation accuracy of the top $u = 10$ best neural networks in 20 generations. The convergence of the evolutionary search algorithm can be observed by the congested distribution and the steadiness of the best architecture. (**b**) The diversity and average fitness increase due to the adjustment of the selection probability. Three different weights of diversity are considered

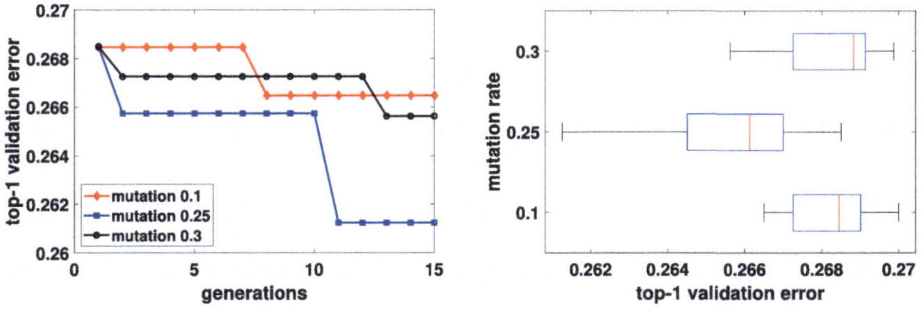

(a) Best models through generations (b) Top-10 models in the final generation

Fig. 8.3 The impact of mutation rate on the evolutionary search algorithm. (**a**) The validation error of the best models throughout the evolutionary search. Exploration and exploitation are balanced by the mutation rate. (**b**) The distribution of the validation error of the top-10 models at the end of the evolution. Better network architectures, which improve the overall performance of the population, are found with a suitable mutation rate of 0.25

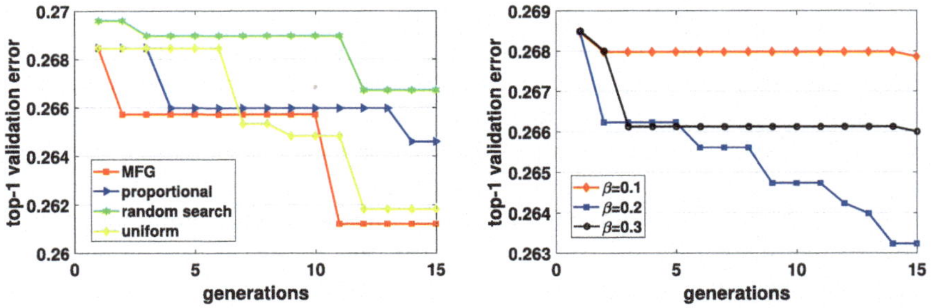

(a) Selection Method Comparison (b) Diversity Comparison

Fig. 8.4 Comparison between different selection methods and diversity. (**a**) Comparison between evolutionary search with proportional selection, uniform selection, and MFG selection. The evolutionary search is performed on the supernet with block design shown in Tables 8.1 and 8.2. The performance of these selection methods is evaluated by the top-1 validation error. (**b**) We compare MFG-based evolutionary search with different population diversity weight β

a random choice block, are examined. For the evolutionary search method, the mutation rate simply represents a trade-off between exploration and exploitation. Because the neural network architecture is mostly defined by the crossing of the selected parents, a lower mutation rate means less exploration and more exploitation. As a result, as seen by the red curve with a mutation rate of 0.1, the search process converged extremely quickly without exploration. In contrast, if the mutation rate is too high, as represented by the black curve, we would lose the advantageous structures developed in prior generations. The blue curve depicts a decent trade-off between exploration and exploitation with a mutation rate of 0.25. This trade-off is further highlighted in Fig. 8.3b, which depicts the population's overall performance when the search process is completed. When the mutation rate is 0.25, a better network design is identified, which enhances the population's overall performance.

In Fig. 8.4b, we depict the validation accuracy of the best architecture of each generation, given different weights on population diversity. When the population is not diverse ($\beta = 0.1$), some architectures with better performance can easily dominate the population at the beginning of the evolution. New architectures mainly come from mutations. They are not guaranteed to improve the performance of the network because they are fully random. Therefore, the validation accuracy of the best architecture remains unchanged since the second generation. When the population is very diverse, we can easily find better architecture at the beginning of the evolution. Therefore, the performance of the best network improves quickly at the beginning. However, those good architectures may not be passed on to the next generations because the weight of the population diversity is very high. Only proper diversity weight ($\beta = 0.2$) can help the population search and preserve new architectures.

Table 8.3 Search cost. Gds–GPU days

Method	Proxyless [21]	SPOS [82]	Ours
Memory cost	37G	24G	32G
Training time	120Gds	96Gds	60Gds
Search time	0	< 1 Gds	< 1 Gds

Table 8.4 Block design details of benchmarks

Schemes	#Blocks	#Choices	Block design
SPOS	20	4	ShuffleNet v2, ShuffleNet xception
FairNAS	19	6	Inverted bottleneck
GreedyNAS	21	7	Inverted bottleneck and identity block

Search Cost Analysis The search cost of the MFG-based NAS algorithm is primarily determined by the training of the supernet and the evaluation of the fitness of the sampled candidates, which necessitates inference of a large number of deep neural networks. Given the block design in Table 8.1 and network design in Table 8.2, the search cost can be illustrated in Table 8.3.

During the evolutionary search, we use dynamic programming to solve the MFG in (8.7). The time complexity of solving the MFG is $O(u^2)$, because we need to construct a $u \times u$ matrix to compute the gradient $\frac{d\rho_i}{dt}$. The time complexity of roulette wheel selection and uniform selection is $O(u)$ and $O(1)$, respectively. Even though the time complexity of the MFG selection probability seems to be higher than other selection methods, its clock time is second-level. To be more precise, given that the $u = 10$, the average running time of the MFG selection is 0.2899 s, which is fairly negligible compared with the training of the supernet and the evaluation of sampled neural networks.

Comparison with State-of-the-Arts By comparing our method with the NAS approaches, shown in Tables 8.4 and 8.5, we conclude that the one-shot NAS with MFG selection has achieved state-of-the-art performance. The best neural architectures for image classification found by three benchmarks with MFG selection mechanisms are illustrated in Fig. 8.5.

We benchmark our approach with three one-shot NAS approaches: SPOS [82], FairNAS [40], and GreedyNAS [203]. The reason is that we use a similar search space design and training method. The designs of the supernet of three benchmarks are shown in Table 8.4. FairNAS uses fairness constraints to ensure equal optimization opportunities for all choice blocks, which fairly estimate the capacity of the choice blocks. They demonstrate that the implementation of the fairness constraints significantly improves the performance of NAS. The search space size of FairNAS is 6^{19}. The blocks of FairNAS are based on an inverted bottleneck. The GreedyNAS greedily shrinks its training space from all architectures into potentially good paths, and further into the candidate pool. The search space of

Table 8.5 Comparison with state-of-the-art NAS approaches. The validation accuracy is top-1 accuracy after search (without fine-tuning). We retrain three benchmarks and validate their accuracy on our own. For other approaches, we use the accuracy reported by the original chapter

Model	FLOPs	Parameters	Top-1 acc(%)
MobileNet V2 [164]	300M	3.4M	72.0
ShuffleNet V2 [141]	299M	3.5M	72.6
DARTS [136]	595M	4.7M	73.1
Proxyless [21]	320M	4.0M	74.2
FBNet [195]	295M	4.5M	74.1
SPOS [82]	323M	3.2 M	73.1
FairNAS [40]	388M	4.6M	67.4
GreedyNAS [203]	320M	3.8M	66.8
One-shot + MFG (ours)	321M	3.4M	74.0

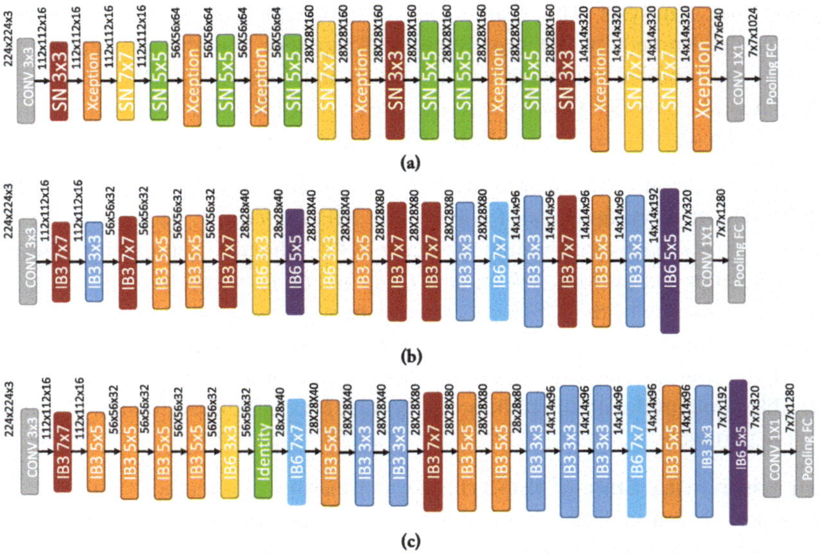

Fig. 8.5 Deep neural network architectures found by three benchmarks with MFG selection mechanism. The grey color is used to show the pre-processing layer, the global pooling layer, and the fully-connected layer, while the choice blocks are denoted by other colors. SN refers to the basic block in the ShuffleNet and IBx refers to the inverted bottleneck with the expansion ratio of 'x'. The height of the rectangle represents the number of input channels. (**a**) Best model found by SPOS with MFG selection. (**b**) Best model found by FairNas with MFG selection. (**c**) Best model found by GreedyNAS with MFG selection

GreedyNAS is 7^{21}. The blocks of GreedyNAS are based on inverted bottleneck and identity blocks. The evaluation of all these approaches is conducted on ImageNet with parameters the same as the original works.

Other NAS approaches use different training methods and search spaces. Therefore, the comparison between them is somewhat unfair. For example, DARTs [136]

are based on the continuous relaxation of the discrete architecture space. In this way, they can allow a more efficient search of architecture using gradient descent. In contrast, we encode the search space in a supernet with discrete choices of architectures. Proxyless [21] also models the NAS as a single training process of the over-parameterized network (similar to the supernet). However, its basic block design is inspired by MobileNet V2 [164] and MnasNet [179].

8.4 Chapter Summary

We propose an evolutionary NAS with MFG selection mechanism. We formulate the selection in the evolutionary search process as an MFG, with the target of maximizing the average fitness and population diversity. Then we use a discrete Fokker-Planck equation to analytically compute the best selection probability for all mating candidates. The selection probability is optimized for each generation, and thus achieves better performance than the widely-used uniform selection and roulette wheel selection, as shown by the simulation results. The proposed one-shot NAS with MFG selection also achieved state-of-the-art performance in the given search space. We also clarify that the proposed MFG formulation is general and can be applied to the search stage for different evolutionary NAS methods, not only limited to the one-shot paradigm in this chapter. Moreover, this MFG formulation can open up additional avenues of research into analytically controlling the evolutionary search process, instead of purely relying on intuition and inspiration from biology.

References

1. Ahuja, S.: Wellposedness of mean field games with common noise under a weak monotonicity condition. SIAM J. Control Optim. **54**(1), 30–48 (2016)
2. Ali-Gombe, A., Elyan, E.: MFC-GAN: class-imbalanced dataset classification using multiple fake class generative adversarial network. Neurocomputing **361**, 212–221 (2019)
3. Alvarez-Melis, D., Jaakkola, T.S.: Gromov-wasserstein alignment of word embedding spaces (2018, preprint). arXiv:1809.00013
4. Amari, S.I., Karakida, R., Oizumi, M.: Information geometry connecting wasserstein distance and kullback–leibler divergence via the entropy-relaxed transportation problem. Inf. Geom. **1**, 13–37 (2018)
5. Antoniou, A., Storkey, A., Edwards, H.: Data augmentation generative adversarial networks (2017, preprint). arXiv:1711.04340
6. Arjovsky, M., Bottou, L.: Towards principled methods for training generative adversarial networks. In: Proceedings of International Conference on Learning Representations. OpenReview.net, Toulon (2017)
7. Arjovsky, M., Chintala, S., Bottou, L.: Wasserstein generative adversarial networks. In: Precup, D., Teh, Y.W. (eds.) Proceedings of Machine Learning Research, vol. 70, pp. 214–223. PMLR, International Convention Centre, Sydney (2017)
8. Arjovsky, M., Chintala, S., Bottou, L.: Wasserstein generative adversarial networks. In: International Conference on Machine Learning, Sydney (2017)
9. Arti, M.: Channel estimation and detection in satellite communication systems. IEEE Trans. Veh. Technol. **65**(12), 10173–10179 (2016)
10. Başar, T., Olsder, G.J.: Dynamic Noncooperative Game Theory. SIAM, Philadelphia (1998)
11. Bauso, D., Tembine, H., Başar, T.: Opinion dynamics in social networks through mean-field games. SIAM J. Control Optim. **54**(6), 3225–3257 (2016)
12. Bellman, R.: Dynamic programming. Science **153**(3731), 34–37 (1966)
13. Benamou, J.D., Carlier, G., Santambrogio, F.: Variational Mean Field Games. Springer, Berlin (2017)
14. Bengio, Y., Mesnil, G., Dauphin, Y., Rifai, S.: Better mixing via deep representations. In: International Conference on Machine Learning, Atlanta (2013)
15. Bensoussan, A., Frehse, J., Yam, P., et al.: Mean Field Games and Mean Field Type Control Theory, vol. 101. Springer, Berlin (2013)
16. Beyer, H.G., Schwefel, H.P.: Evolution strategies–a comprehensive introduction. Nat. Comput. **1**(1), 3–52 (2002)

Y. Kang et al., *Mean Field Guided Machine Learning*, Wireless Networks,
https://doi.org/10.1007/978-3-031-91859-9

17. Bojarski, M., Del Testa, D., Dworakowski, D., Firner, B., Flepp, B., Goyal, P., Jackel, L.D., Monfort, M., Muller, U., Zhang, J., et al.: End to end learning for self-driving cars (2016, preprint). arXiv:1604.07316
18. Bonawitz, K., Ivanov, V., Kreuter, B., Marcedone, A., McMahan, H.B., Patel, S., Ramage, D., Segal, A., Seth, K.: Practical secure aggregation for privacy-preserving machine learning. In: Proceedings of the ACM SIGSAC Conference on Computer and Communications Security (CCS), pp. 1175–1191. Association for Computing Machinery, Dallas (2017)
19. Bonawitz, K., Eichner, H., Grieskamp, W., Huba, D., Ingerman, A., Ivanov, V., Kiddon, C., Konečný, J., Mazzocchi, S., McMahan, B., et al.: Towards federated learning at scale: System design. Proc. Mach. Learn. Syst. **1**(1), 374–388 (2019)
20. Busoniu, L., Babuska, R., De Schutter, B.: A comprehensive survey of multiagent reinforcement learning. IEEE Trans. Syst. Man Cybern. C Appl. Rev. **38**(2), 156–172 (2008)
21. Cai, H., Zhu, L., Han, S.: Proxylessnas: direct neural architecture search on target task and hardware. In: Proceedings of 7th International Conference on Learning Representations, New Orleans (2019)
22. Caines, P.E.: Mean field games. In: Encyclopedia of Systems and Control, pp. 1197–1202 (2021)
23. Caines, P.E., Huang, M., Malhamé, R.P.: Mean field games. In: Handbook of Dynamic Game Theory, vol. 1, pp. 345–372. Springer, Berlin (2018)
24. Cakaj, S.: The parameters comparison of the "starlink" LEO satellites constellation for different orbital shells. Front. Commun. Netw. **2**, 643095 (2021)
25. Cao, H., Guo, X., Laurière, M.: Connecting GANs and MFGs. arXiv preprint (2020)
26. Carmona, R.: Applications of mean field games in financial engineering and economic theory (2020, preprint). arXiv:2012.05237
27. Carmona, R., Delarue, F.: Probabilistic Theory of Mean Field Games with Applications I–II. Springer, Berlin (2018)
28. Carrasco-Casado, A., Do, P.X., Kolev, D., Hosonuma, T., Shiratama, K., Kunimori, H., Trinh, P.V., Abe, Y., Nakasuka, S., Toyoshima, M.: Intersatellite-link demonstration mission between CubeSOTA (LEO CubeSat) and ETS9-HICALI (GEO Satellite). In: IEEE International Conference on Space Optical Systems and Applications (ICSOS), Portland (2019)
29. Chen, Y.C., Zhu, X., Zheng, W.S., Lai, J.H.: Person re-identification by camera correlation aware feature augmentation. IEEE Trans. Pattern Anal. Mach. Intell. **40**(2), 392–408 (2018)
30. Chen, Z., Fu, Y., Zhang, Y., Jiang, Y.G., Xue, X., Sigal, L.: Multi-level semantic feature augmentation for one-shot learning. IEEE Trans. Image Process. **28**(9), 4594–4605 (2019)
31. Chen, P., Liu, S., Zhao, H., Jia, J.: Gridmask data augmentation (2020, preprint). arXiv:2001.04086
32. Chen, S., Liang, Y.C., Sun, S., Kang, S., Cheng, W., Peng, M.: Vision, requirements, and technology trend of 6G: how to tackle the challenges of system coverage, capacity, user data-rate and movement speed. IEEE Wireless Commun. **27**(2), 218–228 (2020)
33. Chen, M., Shlezinger, N., Poor, H.V., Eldar, Y.C., Cui, S.: Communication-efficient federated learning. Proc. Natl. Acad. Sci. USA **118**(17), e2024789118 (2021)
34. Chen, T., Cheng, Y., Gan, Z., Wang, J., Wang, L., Wang, Z., Liu, J.: Adversarial feature augmentation and normalization for visual recognition (2021, preprint). arXiv:2103.12171
35. Chen, X., Zhou, Y., Wu, D., Zhang, W., Zhou, Y., Li, B., Wang, W.: Imagine by reasoning: a reasoning-based implicit semantic data augmentation for long-tailed classification. In: Proceedings of the AAAI Conference on Artificial Intelligence, vol. 36, pp. 356–364 (2022)
36. Cho, Y.J., Wang, J., Joshi, G.: Client selection in federated learning: convergence analysis and power-of-choice selection strategies (2020, preprint). arXiv:2010.01243
37. Chow, S., Li, W., Zhou, H.: Entropy dissipation of Fokker-Planck equations on graphs. Discrete Contin. Dyn. Syst. - A **38**(10), 4929–4950 (2018). https://doi.org/10.3934/dcds.2018215
38. Chow, S.N., Li, W., Lu, J., Zhou, H.: Population games and discrete optimal transport. J. Nonlinear Sci. **29**, 871–896 (2018)

39. Chow, S.N., Dieci, L., Li, W., Zhou, H.: Entropy dissipation semi-discretization schemes for fokker–planck equations. J. Dyn. Differ. Equ. **31**(2), 765–792 (2019)
40. Chu, X., Zhang, B., Xu, R.: Fairnas: rethinking evaluation fairness of weight sharing neural architecture search. In: Proceedings of IEEE/CVF International Conference on Computer Vision Workshop, pp. 12239–12248, Montréal (2021)
41. Ciregan, D., Meier, U., Schmidhuber, J.: Multi-column deep neural networks for image classification. In: 2012 IEEE Conference on Computer Vision and Pattern Recognition Providence (2012)
42. Cousin, A., Crépey, S., Guéant, O., Hobson, D., Jeanblanc, M., Lasry, J.M., Laurent, J.P., Lions, P.L., Tankov, P., Guéant, O., et al.: Mean Field Games and Applications. Springer, Berlin (2011)
43. Creswell, A., White, T., Dumoulin, V., Arulkumaran, K., Sengupta, B., Bharath, A.A.: Generative adversarial networks: an overview. IEEE Signal Process. Mag. **35**(1), 53–65 (2018)
44. Cubuk, E.D., Zoph, B., Mane, D., Vasudevan, V., Le, Q.V.: Autoaugment: learning augmentation policies from data (2018, preprint). arXiv:1805.09501
45. Cubuk, E.D., Zoph, B., Shlens, J., Le, Q.V.: Randaugment: practical automated data augmentation with a reduced search space. In: Proceedings of the IEEE Conference on Computer Vision and Pattern Recognition Workshops (2020)
46. Cui, X., Zhang, W., Tüske, Z., Picheny, M.: Evolutionary stochastic gradient descent for optimization of deep neural networks. In: Proceedings of 32nd International Conference on Neural Information Processing Systems, Montréal (2018)
47. Cunningham, P., Cord, M., Delany, S.J.: Supervised learning. In: Machine Learning Techniques for Multimedia: Case Studies on Organization and Retrieval, pp. 21–49. Springer, Berlin (2008)
48. Dabouei, A., Soleymani, S., Taherkhani, F., Nasrabadi, N.M.: Supermix: supervising the mixing data augmentation. In: Proceedings of IEEE/CVF Conference on Computer Vision and Pattern Recognition, pp. 13794–13803 (2021)
49. De Jong, K.A.: An Analysis of the Behavior of a Class of Genetic Adaptive Systems. University of Michigan, Ann Arbor (1975)
50. De Sanctis, M., Cianca, E., Araniti, G., Bisio, I., Prasad, R.: Satellite communications supporting Internet of remote things. IEEE Internet Things J. **3**(1), 113–123 (2015)
51. Deng, L.: The MNIST database of handwritten digit images for machine learning research [best of the web]. IEEE Signal Process. Mag. **29**(6), 141–142 (2012)
52. Deng, J., Dong, W., Socher, R., Li, L.J., Li, K., Fei-Fei, L.: Imagenet: a large-scale hierarchical image database. In: 2009 IEEE Conference on Computer Vision and Pattern Recognition, Miami (2009)
53. Devlin, J., Chang, M.W., Lee, K., Toutanova, K.: Bert: pre-training of deep bidirectional transformers for language understanding (2018, preprint). arXiv:1810.04805
54. DeVries, T., Taylor, G.W.: Dataset augmentation in feature space (2017, preprint). arXiv:1702.05538
55. DeVries, T., Taylor, G.W.: Improved regularization of convolutional neural networks with cutout (2017, preprint). arXiv:1708.04552
56. Dinh, C.T., Tran, N.H., Nguyen, M.N., Hong, C.S., Bao, W., Zomaya, A.Y., Gramoli, V.: Federated learning over wireless networks: convergence analysis and resource allocation. IEEE/ACM Trans. Netw. **29**(1), 398–409 (2020)
57. Douzas, G., Bacao, F.: Effective data generation for imbalanced learning using conditional generative adversarial networks. Expert Syst. Appl. **91**, 464–471 (2018)
58. Dvornik, N., Mairal, J., Schmid, C.: Modeling visual context is key to augmenting object detection datasets. In: Proceedings of European Conference on Computer Vision, Munich (2018)
59. Dwibedi, D., Misra, I., Hebert, M.: Cut, paste and learn: surprisingly easy synthesis for instance detection. In: Proceedings of IEEE International Conference on Computer Vision, Venice (2017)

60. Elbert, B.R.: Introduction to Satellite Communication. Artech House, Norwood (2008)
61. Elfwing, S., Uchibe, E., Doya, K.: Sigmoid-weighted linear units for neural network function approximation in reinforcement learning. Neural Netw. **107**, 3–11 (2018)
62. Elloumi, M., Dhaou, R., Escrig, B., Idoudi, H., Saidane, L.A.: Monitoring road traffic with a UAV-based system. In: IEEE Wireless Communications and Networking Conference, Barcelona (2018)
63. Esteva, A., Kuprel, B., Novoa, R.A., Ko, J., Swetter, S.M., Blau, H.M., Thrun, S.: Dermatologist-level classification of skin cancer with deep neural networks. Nature **542**(7639), 115–118 (2017)
64. Ferradans, S., Papadakis, N., Peyré, G., Aujol, J.F.: Regularized discrete optimal transport. SIAM J. Imag. Sci. **7**(3), 1853–1882 (2014)
65. Flamary, R., Cuturi, M., Courty, N., Rakotomamonjy, A.: Wasserstein discriminant analysis. Mach. Learn. **107**(12), 1923–1945 (2018)
66. Foerster, J., Nardelli, N., Farquhar, G., Afouras, T., Torr, P.H., Whiteson, S.: Stabilising experience replay for deep multi-agent reinforcement learning. In: International Conference on Machine Learning, Vienna, pp. 1438–1447 (2018)
67. Frid-Adar, M., Klang, E., Amitai, M., Goldberger, J., Greenspan, H.: Synthetic data augmentation using gan for improved liver lesion classification. In: IEEE International Symposium on Biomedical (ISBI 2018), Washington (2018)
68. Friedkin, N., Johnsen, E.: Social influence networks and opinion change. Adv. Group Process. **16**(1), 1–29 (1999)
69. Gao, H., Li, W., Banez, R.A., Han, Z., Poor, H.V.: Mean field evolutionary dynamics in ultra dense mobile edge computing systems. In: Proceedings of IEEE Global Communications Conference, Waikoloa (2019)
70. Gao, H., Lee, W., Li, W., Han, Z., Osher, S., Poor, H.V.: Energy-efficient velocity control for massive rotary-wing UAVs: a mean field game approach. In: Proceedings of IEEE Global Communications Conference, Taipei (2020)
71. Gao, H., Li, W., Banez, R.A., Han, Z., Poor, H.V.: Mean field evolutionary dynamics in dense-user multi-access edge computing systems. IEEE Trans. Wireless Commun. **19**(12), 7825–7835 (2020). https://doi.org/10.1109/TWC.2020.3016695
72. Gao, Z., Wu, Y., Jia, Y., Harandi, M.: Hyperbolic feature augmentation via distribution estimation and infinite sampling on manifolds. Adv. Neural Inf. Process. Syst. **35**, 34421–34435 (2022)
73. Georgakis, G., Mousavian, A., Berg, A.C., Kosecka, J.: Synthesizing training data for object detection in indoor scenes (2017, preprint). arXiv:1702.07836
74. Ghiasi, G., Cui, Y., Srinivas, A., Qian, R., Lin, T.Y., Cubuk, E.D., Le, Q.V., Zoph, B.: Simple copy-paste is a strong data augmentation method for instance segmentation. In: Proceedings of IEEE/CVF Conference on Computer Vision and Pattern Recognition, Nashville (2021)
75. Glaser, P.E.: An overview of the solar power satellite option. IEEE Trans. Microw. Theory Tech. **40**(6), 1230–1238 (1992)
76. Goh, G.B., Siegel, C., Vishnu, A., Hodas, N.O., Baker, N.A.: Deep learning for computational chemistry. J. Comput. Chem. **38**(16), 1291–1307 (2017)
77. Goldberg, D.E.: Genetic Algorithms in Search, Optimization and Machine Learning, 1st edn. Addison-Wesley, Boston (1989)
78. Gong, C., Wang, D., Li, M., Chandra, V., Liu, Q.: Keepaugment: a simple information-preserving data augmentation approach. In: Proceedings of IEEE/CVF Conference on Computer Vision and Pattern Recognition (2021)
79. Gontijo-Lopes, R., Smullin, S.J., Cubuk, E.D., Dyer, E.: Affinity and diversity: quantifying mechanisms of data augmentation (2020, preprint). arXiv:2002.08973
80. Goodfellow, I., Pouget-Abadie, J., Mirza, M., Xu, B., Warde-Farley, D., Ozair, S., Courville, A., Bengio, Y.: Generative adversarial nets. In: Advances in Neural Information Processing Systems, pp. 2672–2680. Montreal (2014)
81. Goodfellow, I., Pouget-Abadie, J., Mirza, M., Xu, B., Warde-Farley, D., Ozair, S., Courville, A., Bengio, Y.: Generative adversarial nets. Adv. Neural Inf. Process. Syst. **27**, 2672–2680 (2014)

82. Guo, Z., Zhang, X., Mu, H., Heng, W., Liu, Z., Wei, Y., Sun, J.: Single path one-shot neural architecture search with uniform sampling. In: Proceedings of 16th European Conference on Computer Vision, Glasgow, vol. 17, pp. 544–560 (2020)
83. Haddadpour, F., Mahdavi, M.: On the convergence of local descent methods in federated learning (2019, preprint). arXiv:1910.14425
84. Harris, E., Marcu, A., Painter, M., Niranjan, M., Prügel-Bennett, A., Hare, J.: Fmix: enhancing mixed sample data augmentation (2020, preprint). arXiv:2002.12047
85. He, K., Zhang, X., Ren, S., Sun, J.: Deep residual learning for image recognition. In: Proceedings of the IEEE Conference on Computer Vision and Pattern Recognition, Las Vegas (2016)
86. He, X., Zhao, K., Chu, X.: Automl: a survey of the state-of-the-art. Knowl.-Based Syst. **212**, 106622–106649 (2021)
87. Hegselmann, R., Krause, U.: Opinion dynamics and bounded confidence models, analysis and simulation. J. Artif. Soc. Soc. Simul. **5**(3), 1–33 (2002)
88. Hendrycks, D., Mu, N., Cubuk, E.D., Zoph, B., Gilmer, J., Lakshminarayanan, B.: Augmix: a simple data processing method to improve robustness and uncertainty (2019, preprint). arXiv:1912.02781
89. Ho, D., Liang, E., Chen, X., Stoica, I., Abbeel, P.: Population based augmentation: efficient learning of augmentation policy schedules. In: International Conference on Machine Learning, Long Beach (2019)
90. Holt, C.A., Roth, A.E.: The Nash equilibrium: a perspective. Proc. Natl. Acad. Sci. **101**(12), 3999–4002 (2004)
91. Hu, J., Wellman, M.P.: Multiagent reinforcement learning: theoretical framework and an algorithm. In: Proceedings of International Conference on Machine Learning (ICML), Madison (1998)
92. Hu, B., Gao, Y., Liu, L., Ma, H.: Federated region-learning: an edge computing based framework for urban environment sensing. In: IEEE Global Communications Conference, Abu Dhabi (2018)
93. Huang, M., Malhamé, R.P., Caines, P.E.: Large population stochastic dynamic games: closed-loop McKean-Vlasov systems and the Nash certainty equivalence principle. Commun. Inf. Syst. Commun. Inf. Syst. **6**(3), 221–252 (2006)
94. Huang, S.W., Lin, C.T., Chen, S.P., Wu, Y.Y., Hsu, P.H., Lai, S.H.: AugGAN: cross domain adaptation with gan-based data augmentation. In: Proceedings of European Conference on Computer Vision, Munich (2018)
95. Imteaj, A., Thakker, U., Wang, S., Li, J., Amini, M.H.: A survey on federated learning for resource-constrained IOT devices. IEEE Internet Things J. **9**(1), 1–24 (2021)
96. Inoue, H.: Data augmentation by pairing samples for images classification (2018, preprint). arXiv:1801.02929
97. Jackson, P.T., Abarghouei, A.A., Bonner, S., Breckon, T.P., Obara, B.: Style augmentation: data augmentation via style randomization. In: CVPR Workshops, Long Beach (2019)
98. Jacobs, M., Léger, F., Li, W., Osher, S.: Solving large-scale optimization problems with a convergence rate independent of grid size. SIAM J. Numer. Anal. **57**(3), 1100–1123 (2019)
99. Jaiswal, A., Babu, A.R., Zadeh, M.Z., Banerjee, D., Makedon, F.: A survey on contrastive self-supervised learning. Technologies **9**(1), 2 (2020)
100. Jin, H., Chollet, F., Song, Q., Hu, X.: Autokeras: an automl library for deep learning. J. Mach. Learn. Res. **24**(6), 1–6 (2023)
101. Jing, Y., Jiang, C., Ge, N., Kuang, L.: Resource optimization for signal recognition in satellite MEC with federated learning. In: International Conference on Wireless Communications and Signal Processing, Changsha (2021)
102. Kairouz, P., McMahan, H.B., Avent, B., Bellet, A., Bennis, M., Bhagoji, A.N., et al.: Advances and open problems in federated learning (2019, preprint). arXiv:1912.04977

103. Kang, Y., Liu, S., Zhang, H., Li, W., Han, Z., Osher, S., Poor, H.V.: Joint sensing task assignment and collision-free trajectory optimization for mobile vehicle networks using mean-field games. IEEE Internet Things J. **8**(10), 8488–8503 (2020)
104. Kang, Y., Liu, S., Zhang, H., Han, Z., Osher, S., Poor, H.V.: Task selection and collision-free route planning for mobile crowdsensing using multi-population mean-field games. IEEE Trans. Green Commun. Netw. **5**(4), 1947–1960 (2021)
105. Kang, Y., Wang, H., Kim, B., Xie, J., Zhang, X.P., Han, Z.: Time efficient offloading optimization in automotive multi-access edge computing networks using mean-field games. IEEE Trans. Veh. Technol. **72**(5), 6460–6473 (2023)
106. Khan, L.U., Pandey, S.R., Tran, N.H., Saad, W., Han, Z., Nguyen, M.N., Hong, C.S.: Federated learning for edge networks: resource optimization and incentive mechanism. IEEE Commun. Mag. **58**(10), 88–93 (2020)
107. Khedo, K.K., Perseedoss, R., Mungur, A., et al.: A wireless sensor network air pollution monitoring system (2010, preprint). arXiv:1005.1737
108. Kim, J.H., Choo, W., Song, H.O.: Puzzle mix: exploiting saliency and local statistics for optimal mixup. In: International Conference on Machine Learning, Vienna (2020)
109. Kim, J.H., Choo, W., Jeong, H., Song, H.O.: Co-mixup: saliency guided joint mixup with supermodular diversity (2021, preprint). arXiv:2102.03065
110. Kingma, D.P., Welling, M.: Auto-encoding variational bayes (2013, preprint). arXiv:1312.6114
111. Kiran, B.R., Sobh, I., Talpaert, V., Mannion, P., Al Sallab, A.A., Yogamani, S., Pérez, P.: Deep reinforcement learning for autonomous driving: a survey. IEEE Trans. Intell. Transp. Syst. **23**(6), 4909–4926 (2021)
112. Kober, J., Bagnell, J.A., Peters, J.: Reinforcement learning in robotics: a survey. Int. J. Robot. Res. **32**(11), 1238–1274 (2013)
113. Kodheli, O., Lagunas, E., Maturo, N., Sharma, S.K., Shankar, B., Montoya, J.F.M., Duncan, J.C.M., Spano, D., Chatzinotas, S., Kisseleff, S., et al.: Satellite communications in the new space era: a survey and future challenges. IEEE Commun. Surv. Tuts. **23**(1), 70–109 (2020)
114. Krizhevsky, A., Hinton, G., et al.: Learning multiple layers of features from tiny images. In: Citeseer, Toronto (2009)
115. Kumar, V., Bhaskaran, V., Mirchandani, R.: Customer segmentation and strategy development based on customer lifetime value: a case study. Expert Syst. Appl. **61**(1), 206–216 (2016)
116. Kuo, C.W., Ma, C.Y., Huang, J.B., Kira, Z.: Featmatch: feature-based augmentation for semi-supervised learning. In: European Conference on Computer Vision, Glasgow (2020)
117. Lasry, J.M., Lions, P.L.: Mean field games. Japanese J. Math. **2**(1), 229–260 (2007)
118. LeDell, E., Poirier, S.: H2o automl: scalable automatic machine learning. In: Proceedings of AutoML Workshop at ICML, San Diego (2020)
119. Lee, D., Park, H., Pham, T., Yoo, C.D.: Learning augmentation network via influence functions. In: Proceedings of IEEE/CVF Conference on Computer Vision and Pattern Recognition, Seattle (2020)
120. Lee, J., Kim, E., Lee, J., Lee, J., Choo, J.: Learning debiased representation via disentangled feature augmentation. Adv. Neural Inf. Process. Syst. **34**, 25123–25133 (2021)
121. Lee, W., Liu, S., Tembine, H., Li, W., Osher, S.: Controlling propagation of epidemics via mean-field control. SIAM J. Appl. Math. **81**(1), 190–207 (2021)
122. Lewis, F.L., Vrabie, D., Syrmos, V.L.: Optimal Control. Wiley, Hoboken (2012)
123. Li, W., Yin, P., Osher, S.: Computations of optimal transport distance with Fisher information regularization. J. Sci. Comput. **75**(3), 1581–1595 (2018)
124. Li, H., Zhang, X., Tian, Q., Xiong, H.: Attribute mix: semantic data augmentation for fine grained recognition. In: IEEE International Conference on Visual Communications and Image Processing, Macau (2020)
125. Li, L., Cheng, Q., Tang, X., Bai, T., Chen, W., Ding, Z., Han, Z.: Resource allocation for NOMA-MEC systems in ultra-dense networks: a learning aided mean-field game approach. IEEE Trans. Wireless Commun. **20**(3), 1487–1500 (2020)

126. Li, T., Sahu, A.K., Talwalkar, A., Smith, V.: Federated learning: challenges, methods, and future directions. IEEE Signal Process. Mag. **37**(3), 50–60 (2020)
127. Li, B., Wu, F., Lim, S.N., Belongie, S., Weinberger, K.Q.: On feature normalization and data augmentation. In: Proceedings of IEEE/CVF Conference on Computer Vision and Pattern Recognition (2021)
128. Li, P., Li, D., Li, W., Gong, S., Fu, Y., Hospedales, T.M.: A simple feature augmentation for domain generalization. In: Proceedings of IEEE/CVF International Conference on Computer Vision, Montreal (2021)
129. Li, P., Liu, X., Xie, X.: Learning sample-specific policies for sequential image augmentation. In: Proceedings of ACM International Conference on Multimedia, Nice (2021)
130. Lillicrap, T.P., Hunt, J.J., Pritzel, A., Heess, N., Erez, T., Tassa, Y., Silver, D., Wierstra, D.: Continuous control with deep reinforcement learning (2015, preprint). arXiv:1509.02971
131. Lim, W.Y.B., Ng, J.S., Xiong, Z., Jin, J., Zhang, Y., Niyato, D., Leung, C., Miao, C.: Decentralized edge intelligence: a dynamic resource allocation framework for hierarchical federated learning. IEEE Trans. Parallel Distrib. Syst. **33**(3), 536–550 (2021)
132. Lin, A.T., Fung, S.W., Li, W., Nurbekyan, L., Osher, S.J.: APAC-net: alternating the population and agent control via two neural networks to solve high-dimensional stochastic mean field games (2020, preprint). arXiv:2002.10113
133. Lin, A.T., Fung, S.W., Li, W., Nurbekyan, L., Osher, S.J.: Alternating the population and control neural networks to solve high-dimensional stochastic mean-field games. Proc. Natl. Acad. Sci. U.S.A. **118**(31), e2024713118 (2021)
134. Liu, M.Y., Tuzel, O.: Coupled generative adversarial networks. In: Advances in Neural Inform. Processing System, Barcelona (2016)
135. Liu, B., Wang, X., Dixit, M., Kwitt, R., Vasconcelos, N.: Feature space transfer for data augmentation. In: Proceedings of the IEEE Computer Society Conference on Computer Vision and Pattern Recognition, Salt Lake City, pp. 9090–9098 (2018)
136. Liu, H., Simonyan, K., Yang, Y.: DARTS: differentiable architecture search. In: Proceedings of the 7th International Conference on Learning Representations, New Orleans (2019)
137. Liu, X., Zhang, F., Hou, Z., Mian, L., Wang, Z., Zhang, J., Tang, J.: Self-supervised learning: generative or contrastive. IEEE Trans. Knowl. Data Eng. **35**(1), 857–876 (2021)
138. Liu, J., Chai, C., Luo, Y., Lou, Y., Feng, J., Tang, N.: Feature augmentation with reinforcement learning. In: IEEE International Conference on Data Engineering, Kuala Lumpur (2022)
139. Loshchilov, I., Hutter, F.: SGDR: stochastic gradient descent with warm restarts (2016, preprint). arXiv:1608.03983
140. Lowe, R., Wu, Y., Tamar, A., Harb, J., Abbeel, P., Mordatch, I.: Multi-agent actor-critic for mixed cooperative-competitive environments. In: Advances in Neural Information Processing Systems, Long Beach, pp. 6379–6390 (2017)
141. Ma, N., Zhang, X., Zheng, H.T., Sun, J.: Shufflenet v2: practical guidelines for efficient CNN architecture design. In: Proceedings of European Conference on Computer Vision (ECCV), Munich, vol. 11218, pp. 116–131 (2018)
142. Madan, A., Cebrian, M., Moturu, S., Farrahi, K., Pentland, A.: Sensing the health state of a community. IEEE Pervasive Comput. **11**(4), 36–45 (2012)
143. Mariani, G., Scheidegger, F., Istrate, R., Bekas, C., Malossi, C.: Bagan: data augmentation with balancing gan (2018, preprint). arXiv:1803.09655
144. Masisi, L., Nelwamondo, V., Marwala, T.: The use of entropy to measure structural diversity. In: IEEE International Conference on Computational Cybernetics, Stara Lesna, pp. 41–45 (2008)
145. McMahan, B., Moore, E., Ramage, D., Hampson, S., y Arcas, B.A.: Communication-efficient learning of deep networks from decentralized data. In: Artificial Intelligence and Statistics, Fort Lauderdale (2017)
146. McMahan, B., Moore, E., Ramage, D., Hampson, S., y Arcas, B.A.: Communication-efficient learning of deep networks from decentralized data. In: Proceedings of Artificial Intelligence and Statistics, pp. 1273–1282. Fort Lauderdale (2017)

147. Mikołajczyk, A., Grochowski, M.: Data augmentation for improving deep learning in image classification problem. In: International Interdisciplinary PhD Workshop (IIPhDW), Poland, pp. 117–122 (2018)
148. Mnih, V., Kavukcuoglu, K., Silver, D., Rusu, A.A., Veness, J., Bellemare, M.G., Graves, A., Riedmiller, M., Fidjeland, A.K., Ostrovski, G.: Human-level control through deep reinforcement learning. Nature **518**(7540), 529–533 (2015)
149. Ni, W., Liu, Y., Yang, Z., Tian, H., Shen, X.: Federated learning in multi-RIS-aided systems. IEEE Internet Things J. **9**(12), 9608–9624 (2021)
150. Ni, W., Liu, Y., Yang, Z., Tian, H., Shen, X.: Integrating over-the-air federated learning and non-orthogonal multiple access: what role can RIS play? IEEE Trans. Wirel. Commun **21**(12), 10083–10099 (2022)
151. Ni, W., Zheng, J., Tian, H.: Semi-federated learning for collaborative intelligence in massive IOT networks. IEEE Internet Things J. **10**(13), 11942–11943 (2023)
152. Nishizaki, H.: Data augmentation and feature extraction using variational autoencoder for acoustic modeling. In: 2017 Asia-Pacific Signal and Information Processing Association Annual Summit Conference (APSIPA ASC), Kuala Lumpur (2017)
153. Olsson, V., Tranheden, W., Pinto, J., Svensson, L.: Classmix: segmentation-based data augmentation for semi-supervised learning. In: IEEE Winter Conference on Applications of Computer Vision, Waikoloa (2021)
154. Ozair, S., Bengio, Y.: Deep directed generative autoencoders (2014, preprint). arXiv:1410.0630
155. Park, D.S., Zhang, Y., Jia, Y., Han, W., Chiu, C.C., Li, B., Wu, Y., Le, Q.V.: Improved noisy student training for automatic speech recognition (2020, preprint). arXiv:2005.09629
156. Paty, F.P., Cuturi, M.: Subspace robust wasserstein distances. In: International Conference on Machine Learning. Long Beach (2019)
157. Phua, C., Lee, V., Smith-Miles, K., Gayler, R.: A comprehensive survey of data mining-based fraud detection research. Artif. Intell. Rev. **33**(4), 213–238 (2010)
158. Pokhrel, S.R., Choi, J.: Federated learning with blockchain for autonomous vehicles: analysis and design challenges. IEEE Trans. Commun. **68**(8), 4734–4746 (2020)
159. Ratner, A.J., Ehrenberg, H., Hussain, Z., Dunnmon, J., Ré, C.: Learning to compose domain-specific transformations for data augmentation. Adv. Neural Inf. Process. Syst. **30**, 3236–3246 (2017)
160. Rieke, N., Hancox, J., Li, W., Milletari, F., Roth, H.R., Albarqouni, S., Bakas, S., Galtier, M.N., Landman, B.A., Maier-Hein, K., et al.: The future of digital health with federated learning. NPJ Digit. Med. **3**(1), 119 (2020)
161. Russakovsky, O., Deng, J., Su, H., Krause, J., Satheesh, S., Ma, S., Huang, Z., Karpathy, A., Khosla, A., Bernstein, M., et al.: Imagenet large scale visual recognition challenge. Int. J. Comput. Vis. **115**(3), 211–252 (2015)
162. Sandfort, V., Yan, K., Pickhardt, P.J., Summers, R.M.: Data augmentation using generative adversarial networks (cyclegan) to improve generalizability in CT segmentation tasks. Sci. Rep. **9**(1), 16884 (2019)
163. Sandholm, W.H.: Pairwise comparison dynamics and evolutionary foundations for nash equilibrium. Games **1**(1), 3–17 (2009)
164. Sandler, M., Howard, A., Zhu, M., Zhmoginov, A., Chen, L.C.: Mobilenetv2: inverted residuals and linear bottlenecks. In: Proceedings of IEEE Conference on Computer Vision and Pattern Recognition, Salt Lake City, pp. 4510–4520 (2018)
165. Sarikaya, Y., Ercetin, O.: Motivating workers in federated learning: a Stackelberg game perspective. IEEE Netw. Lett. **2**(1), 23–27 (2019)
166. Sato, I., Nishimura, H., Yokoi, K.: APAC: augmented pattern classification with neural networks (2015, preprint). arXiv:1505.03229
167. Schulte, J.: Adjoint methods for hamilton-jacobi-bellman equations. Diploma Thesis (2010)
168. Shalev-Shwartz, S., Shammah, S., Shashua, A.: On a formal model of safe and scalable self-driving cars. In: Proceedings of 1st Conference on Robot Learning, Mountain View, vol. 1, pp. 1–16 (2017)

169. Sheetz, M., Petrova, M.: Why in the next decade companies will launch thousands more satellites than in all of history. https://www.cnbc.com/2019/12/14/spacex-oneweb-and-amazon-to-launch-thousands-more-satellites-in-2020s.html
170. Sheller, M.J., Edwards, B., Reina, G.A., Martin, J., Pati, S., Kotrotsou, A., Milchenko, M., Xu, W., Marcus, D., Colen, R.R., et al.: Federated learning in medicine: facilitating multi-institutional collaborations without sharing patient data. Sci. Rep. **10**(1), 12598–12610 (2020)
171. Sherwin, W.B.: Entropy and information approaches to genetic diversity and its expression: genomic geography. Entropy **12**(7), 1765–1798 (2010)
172. Silver, D., Huang, A., Maddison, C.J., Guez, A., Sifre, L., Van Den Driessche, G., Schrittwieser, J., Antonoglou, I., Panneershelvam, V., Lanctot, M.: Mastering the game of go with deep neural networks and tree search. Nature **529**(7587), 484–489 (2016)
173. Simard, P.Y., Steinkraus, D., Platt, J.C.: Best practices for convolutional neural networks applied to visual document analysis. In: Proceedings of the Seventh International Conference on Document Analysis and Recognition, Edinburgh, Scotland, vol. 3, pp. 958–962 (2003)
174. Singh, K.K., Yu, H., Sarmasi, A., Pradeep, G., Lee, Y.J.: Hide-and-seek: a data augmentation technique for weakly-supervised localization and beyond (2018, preprint). arXiv:1811.02545
175. Summers, C., Dinneen, M.J.: Improved mixed-example data augmentation. In: 2019 IEEE Winter Conference on Applications of Computer Vision (WACV), Waikoloa (2019)
176. Sutton, R.S., Barto, A.G.: Reinforcement Learning: An Introduction. MIT Press, Cambridge (2018)
177. Takase, T., Karakida, R., Asoh, H.: Self-paced data augmentation for training neural networks. Neurocomputing **442**, 296–306 (2021)
178. Tan, M., Le, Q.: Efficientnet: rethinking model scaling for convolutional neural networks. In: International Conference on Machine Learning, Long Beach (2019)
179. Tan, M., Chen, B., Pang, R., Vasudevan, V., Sandler, M., Howard, A., Le, Q.V.: Mnasnet: platform-aware neural architecture search for mobile. In: Proceedings of IEEE/CVF Conference on Computer Vision and Pattern Recognition, Long Beach, pp. 2820–2828 (2019)
180. Tanaka, F.H.K.D.S., Aranha, C.: Data augmentation using gans (2019, preprint). arXiv:1904.09135
181. Tang, Z., Peng, X., Li, T., Zhu, Y., Metaxas, D.N.: Adatransform: adaptive data transformation. In: Proceedings of IEEE/CVF International Conference on Computer Vision, Seoul (2019)
182. Taylor, L., Nitschke, G.: Improving deep learning with generic data augmentation. In: 2018 IEEE Symposium Series on Computational Intelligence (SSCI), pp. 1542–1547. Bangalore (2018)
183. Tokozume, Y., Ushiku, Y., Harada, T.: Between-class learning for image classification. In: Proceedings of IEEE Conference on Computer Vision and Pattern Recognition, Salt Lake City, pp. 5486–5494 (2018)
184. Tran, T., Pham, T., Carneiro, G., Palmer, L., Reid, I.: A Bayesian data augmentation approach for learning deep models. Adv. Neural Inf. Process. Syst. **30**, 2797–2806 (2017)
185. Tsantekidis, A., Passalis, N., Tefas, A., Kanniainen, J., Gabbouj, M., Iosifidis, A.: Forecasting stock prices from the limit order book using convolutional neural networks. IEEE Trans. Neural Netw. Learn. Syst. **29**(10), 4801–4812 (2017)
186. Umarov, S., Hahn, M., Kobayashi, K.: Beyond the Triangle: Brownian Motion, Ito Calculus, and Fokker-planck Equation-Fractional Generalizations. World Scientific, Singapore (2018)
187. Vaswani, A., Shazeer, N., Parmar, N., Uszkoreit, J., Jones, L., Gomez, A.N., Kaiser, Ł., Polosukhin, I.: Attention is all you need. In: Advances in Neural Information Processing Systems, Long Beach (2017)
188. Verma, V., Lamb, A., Beckham, C., Najafi, A., Mitliagkas, I., Lopez-Paz, D., Bengio, Y.: Manifold mixup: Better representations by interpolating hidden states. In: International Conference on Machine Learning, Long Beach (2019)
189. Volpi, R., Morerio, P., Savarese, S., Murino, V.: Adversarial feature augmentation for unsupervised domain adaptation. In: Proceedings of IEEE Conference on Computer Vision and Pattern Recognition, Salt Lake City, pp. 5495–5504 (2018)

190. Wang, S., Lee, M., Hosseinalipour, S., Morabito, R., Chiang, M., Brinton, C.G.: Device sampling for heterogeneous federated learning: theory, algorithms, and implementation. In: Proceedings of IEEE Conference on Computer Communications, Virtual (2021)
191. Wang, Y., Huang, G., Song, S., Pan, X., Xia, Y., Wu, C.: Regularizing deep networks with semantic data augmentation. IEEE Trans. Pattern Anal. Mach. Intell. **44**(7), 3733–3748 (2021)
192. Wang, Y., Pan, X., Song, S., Zhang, H., Huang, G., Wu, C.: Implicit semantic data augmentation for deep networks. In: Advances in Neural Information Processing Systems (2019)
193. Weng, L.: From gan to wgan (2019, preprint). arXiv:1904.08994
194. Wong, S.C., Gatt, A., Stamatescu, V., McDonnell, M.D.: Understanding data augmentation for classification: when to warp? In: International Conference on Digital Image Computing: Techniques and Applications (DICTA), Gold Coast (2016)
195. Wu, B., Dai, X., Zhang, P., Wang, Y., Sun, F., Wu, Y., Tian, Y., Vajda, P., Jia, Y., Keutzer, K.: FBNet: hardware-aware efficient convnet design via differentiable neural architecture search. In: Proceedings of IEEE/CVF Conference on Computer Vision and Pattern Recognition, Long Beach, pp. 10734–10742 (2019)
196. Xie, M., Yin, H., Wang, H., Xu, F., Chen, W., Wang, S.: Learning graph-based POI embedding for location-based recommendation. In: Proceedings of the ACM International Conference on Information and Knowledge Management, Indianapolis (2016)
197. Xu, Z., Meng, A., Shi, Z., Yang, W., Chen, Z., Huang, L.: Continuous copy-paste for one-stage multi-object tracking and segmentation. In: Proceedings of IEEE International Conference on Computer Vision, Montreal (2021)
198. Xu, M., Yoon, S., Fuentes, A., Park, D.S.: A comprehensive survey of image augmentation techniques for deep learning. Pattern Recognit. **137**, 109347 (2023)
199. Yang, J., Ye, X., Trivedi, R., Xu, H., Zha, H.: Learning deep mean field games for modeling large population behavior. In: Proceedings of International Conference on Learning Representations (ICLR), Vancouver (2018)
200. Yang, Q., Liu, Y., Chen, T., Tong, Y.: Federated machine learning: concept and applications. ACM Trans. Intell. Syst. Technol. **10**(2), 1–19 (2019)
201. Yang, Z., Chen, M., Saad, W., Hong, C.S., Shikh-Bahaei, M.: Energy efficient federated learning over wireless communication networks. IEEE Trans. Wireless Commun. **20**(3), 1935–1949 (2020)
202. Yang, Z., Chen, M., Wong, K.K., Poor, H.V., Cui, S.: Federated learning for 6G: applications, challenges, and opportunities. Engineering **8**, 33–41 (2022)
203. You, S., Huang, T., Yang, M., Wang, F., Qian, C., Zhang, C.: GreedyNAS: towards fast one-shot NAS with greedy supernet. In: Proceedings of IEEE/CVF Conference on Computer Vision and Pattern Recognition, pp. 1999–2008. Seattle (2020)
204. Yun, S., Han, D., Oh, S.J., Chun, S., Choe, J., Yoo, Y.: Cutmix: regularization strategy to train strong classifiers with localizable features. In: Proceedings of IEEE/CVF International Conference on Computer Vision (2019)
205. Zang, Y., Huang, C., Loy, C.C.: FASA: feature augmentation and sampling adaptation for long-tailed instance segmentation. In: Proceedings of IEEE/CVF International Conference on Computer Vision Montreal (2021)
206. Zeng, R., Zeng, C., Wang, X., Li, B., Chu, X.: A comprehensive survey of incentive mechanism for federated learning (2021, preprint). arXiv:2106.15406
207. Zhan, Y., Zhang, J.: An incentive mechanism design for efficient edge learning by deep reinforcement learning approach. In: Proceedings of IEEE INFOCOM, Toronto, pp. 2489–2498 (2020)
208. Zhan, Y., Li, P., Qu, Z., Zeng, D., Guo, S.: A learning-based incentive mechanism for federated learning. IEEE Internet Things J. **7**(7), 6360–6368 (2020)
209. Zhang, W., Wen, Y., Guan, K., Kilper, D., Luo, H., Wu, D.O.: Energy-optimal mobile cloud computing under stochastic wireless channel. IEEE Trans. Wireless Commun. **12**(9), 4569–4581 (2013)

210. Zhang, H., Cisse, M., Dauphin, Y.N., Lopez-Paz, D.: Mixup: beyond empirical risk minimization (2017, preprint). arXiv:1710.09412
211. Zhang, X., Wang, Q., Zhang, J., Zhong, Z.: Adversarial autoaugment (2019, preprint). arXiv:1912.11188
212. Zhang, H., Kang, Y., Song, L., Han, Z., Poor, H.V.: Age of information minimization for grant-free non-orthogonal massive access using mean-field games. IEEE Trans. Commun. **69**(11), 7806–7820 (2021)
213. Zhang, J., Zhang, Y., Xu, X.: Objectaug: object-level data augmentation for semantic image segmentation. In: International Joint Conference on Neural Networks, Shenzhen (2021)
214. Zhang, Z., Guan, C., Chen, H., Yang, X., Gong, W., Yang, A.: Adaptive privacy-preserving federated learning for fault diagnosis in internet of ships. IEEE Internet Things J. **9**(9), 6844–6854 (2021)
215. Zhang, Y., Zhu, H., Song, Z., Koniusz, P., King, I.: Spectral feature augmentation for graph contrastive learning and beyond. In: Proceedings of the AAAI Conference on Artificial Intelligence, Washington, vol. 37, pp. 11289–11297 (2023)
216. Zheng, G., Zhang, F., Zheng, Z., Xiang, Y., Yuan, N.J., Xie, X., Li, Z.: DRN: a deep reinforcement learning framework for news recommendation. In: Proceedings of 2018 World Wide Web Conference, Lyon, pp. 167–176 (2018)
217. Zhong, Z., Zheng, L., Kang, G., Li, S., Yang, Y.: Random erasing data augmentation. In: Proceedings of the AAAI Conference on Artificial Intelligence, vol. 34, pp. 13001–13008
218. Zhu, Q., Başar, T.: Multi-resolution large population stochastic differential games and their application to demand response management in the smart grid. Dyn. Games Appl. **3**, 68–88 (2013)
219. Zhu, Q., Tembine, H., Başar, T.: Hybrid risk-sensitive mean-field stochastic differential games with application to molecular biology. In: IEEE Conference on Decision and Control and European Control Conference, Orlando, pp. 4491–4497 (2011)
220. Zou, Y., Feng, S., Niyato, D., Jiao, Y., Gong, S., Cheng, W.: Mobile device training strategies in federated learning: an evolutionary game approach. In: 2019 International Conference on Internet of Things (iThings) and IEEE Green Computing and Communications (GreenCom) and IEEE Cyber, Physical and Social Computing (CPSCom) and IEEE Smart Data (SmartData), Atlanta (2019)

Index

Y. Kang et al., *Mean Field Guided Machine Learning*, Wireless Networks, https://doi.org/10.1007/978-3-031-91859-9

The manufacturer's authorised representative in the EU is Springer
Nature Customer Service Centre GmbH, Europaplatz 3, 69115 Heidelberg,
Germany. If you have any concerns regarding our products, please
contact ProductSafety@springernature.com

Printed and bound by CPI Group (UK) Ltd, Croydon, CR0 4YY
30/09/2026
02225607-0013